Human Needs' Analysis and Evaluation Model for Product Development

"The importance of having this book in bibliographic format is reinforced to reach professional and researchers of the field. The information and knowledge generated by this research can reach the niche that may lead to the reflections raised by the authors.

The authors mention new ways of looking at the Product Development Process, identifying gaps and advances, according to the new product production contexts, which must be done constantly. The exponential speed of changes in contexts, technologies, and behaviors requires that academic and professional community constantly review the concepts already established, identifying how they can become appropriate to the times in which they are applied. In addition, any possibility of models that contain dynamism and flexibility can respond more assertively to this conformation, which the authors demonstrate in their work.

The dissertated theme, in addition to being unprecedented, presents practical possibilities of applications in the most diverse design situations and diverse types of products, bringing important insights for the verification of aspects by professionals in the field.

It is a dense and iterative research, which adds knowledge in a forceful way to the community that develops works directed to the human being, especially of products. The quality of the proposed content is emphasized, with a rich and detailed search path in a true and sincere way. In defending the thesis, I stressed the importance of exploring it further, disseminating it in numerous ways, including in book format, so that the knowledge in this field can be advanced.

In this way, I strongly recommend the publication of this content in book format, so that you can achieve your goal of disseminating structured knowledge and assisting professionals and academics in the application of concepts related to human needs for product development."

—Professor Luciane Hilu, Ph.D.

Gabriela Unger Unruh · Osiris Canciglieri Junior

Human Needs' Analysis and Evaluation Model for Product Development

Gabriela Unger Unruh
Fine Arts School (EBA)
Pontifical Catholic University of Parana
Curitiba, Brazil

Osiris Canciglieri Junior
Polytechnic School
Pontifical Catholic University of Parana
Curitiba, Brazil

ISBN 978-3-031-12622-2   ISBN 978-3-031-12623-9 (eBook)
https://doi.org/10.1007/978-3-031-12623-9

Jointly published with Associação Paranaense de Cultura
ISBN of the Co-Publisher's edition: 978-65-5385-032-3

This Springer imprint is published by the registered company Springer Nature Switzerland AG
The registered company address is: Gewerbestrasse 11, 6330 Cham, Switzerland

I would like to dedicate this book to my husband and gratitude for his support and partnership in life and especially in my studies related to this book.

Gabriela Unger Unruh

I would like to dedicate this book especially to my family with affection and gratitude for always being present and participating in my life and especially in my journey as a researcher and Professor.

Osiris Canciglieri Junior

Preface

The essence of product development is to satisfy human needs, in order to improve people's and organization's lives, where product development processes (PDP) systematize each stage, specifying areas involved for the achievement of consistent products with specific development objectives. This process involves multidisciplinary and sociotechnical knowledge, such as human and system aspects. Since the success of the product, and consequently of the company, depends on its ability to satisfy people's needs, human involvement as a decision factor in the process is essential, and for this reason, this book presents a model (HUNE—human needs) that assist in the insertion of human aspects in the product development process (PDP), at the beginning of a design, at the analyze information, during its development and post-development, evaluating its suitability for human beings. The model proved to be actual with respect to the existing ones, dynamic and flexible, because it does not replace any model but can be applied to other models, methods, or structures of PDPs, and enables scope, replication, and future improvements. Its applications brought satisfactory results and it was very well evaluated by the participants in the application, by external experts and through scientific publications. The book was structured into four chapters as following:

(i) Chapter 1: **Introduction**—It presents a general idea of the fields involved in the study and a contextual view of the importance of considering human needs in product development.

(ii) Chapter 2: **Human Needs—Analysis and Evaluation Approach for Product Development Context**—It presents conceptually the human needs approach to support product development process.

(iii) Chapter 3: **Human Needs Model (HUNE)**—It presents conceptually the model HUNE proposed in this book, as well as the sequence of steps for its implementation.

(iv) Chapter 4: **Application of the Human Needs Model (HUNE)**—It presents six detailed application cases of the model in the context of product development process and brings a general discussion about the possibilities of different applications of the model and how future related research indications.

Curitiba, Brazil   Gabriela Unger Unruh
                                                                              Osiris Canciglieri Junior

---

The original version of the book was revised: In Chapters 2 and 3, omitted references have been included. In Chapter 4, new figures have been replaced and unwanted references have been removed. The correction to the book is available at https://doi.org/10.1007/978-3-031-12623-9_5

Acknowledgments

The authors would like to express immensely thank to the experts for their participation in the application process of HUNE model (human needs): Adriano Augusto de Miranda, Alda Yoshi Uemura Reche, Alexandre Minoru Sasaki, Ana Maria Kaiser Cardoso, Athon Francisco Staben de Moura Leite, Bernardo Reisdorfer Leite, Bruna do Valle Turbay, Camilla Buttura Chrusciak, Carlos Eduardo Maran Santos, David Ribeiro Tavares, Eduardo Baade, Flávia Passarelli, Jean Carlo Ferreira Tambosi, Joanine Facioli Urnau, Juliana Ribeiro Olmedo, Kassia Renata da Silva Zanão, Larissa De Oliveira Matia Leite, Marcelo de Moraes Botelho, Marcia Gemari Derenevich, Márcia Regina Cordeiro de Souza, Matheus Beltrame Canciglieri, Michele Marcos de Oliveira, Paulo Iensen Filho, Paulo Henrique Palma Setti, Priscila Remas Cordeiro de Carvalho, Rafaela da Rosa Cardoso Riesemberg, Ricardo Luhm, Rodrigo de Alvarenga, Rogério Mariano, Sabrina Tinfer, Samuel Henrique Werlich, Tânia Agio, Tania Perin, Thiago Augusto Aniceski Cezar, Tiago dos Santos Silva, and Walter Ihlenfeld.

About This Book

The fundamental essence of the product development process is to satisfy human needs, in order to improve people's and organizations' lives, specifying the areas involved to fulfill the design objectives and to systematize the unfolding of each step in a consistent way. This process involves multidisciplinary and sociotechnical knowledge, such as human and system aspects. Since the success of the product, and consequently of the company, depends on its ability to satisfy people's needs, its involvement as a decision factor in the process is essential, and for that reason, this book presents the conceptual process of building the product. HUNE model (human needs), based on the systematic review of literature and analysis of the relevant content developed by Unruh (2020), includes human factors in the product development process. The HUNE model was evaluated by several experts following an iterative process taking into account three cycles: (i) analysis; (ii) design; and (iii) prototyping (application) and evaluation. The result was the maturation of the proposed model that serves to assist in the insertion of human aspects in the product development process (PDP). The model has presented a potential when compared to the existing ones in the literature, dynamic and flexible, since it does not replace the existing models but rather complements the gap found, allowing greater coverage, replication, and future improvements of the PDP. The book presents six cases of applications with promising results. In this way, this book was structured and written with the aim of helping designers and engineers and similar professionals in the product development process with the human being as the main focus. The authors believe that this deeper look at human characteristics can contribute to the evolution of the product development process being increasingly associated with the desires and well-being of its users (human beings).

Contents

About the Authors

Dr. Gabriela Unger Unruh is Associate Professor of Design, Ergonomics, Usability, User Experience (UX), and Cognitive Psychology at Pontifical Catholic University of Paraná in Brazil (PUCPR). She did her master and doctor degree in Systems and Industrial Engineering at PUCPR. She has nine years of experience in usability and user experience, working in the academy and in enterprises of physical and digital product development. Prof. Gabriela Unger Unruh's research has been funded directly by PUCPR, is focused on human-centered design, and her work has led five papers. She oriented two scientific initiation designs and six specialization monographies.

Prof. Osiris Canciglieri Junior (Ph.D.) is Full Professor of Industrial Engineering and Control and Automation Engineering (Mechatronic) at Pontifical Catholic University of Paraná in the Brazil (PUCPR). He is also Managing Director of Industrial and Systems Engineering Graduate Program (PPGEPS). He has more than 30 years of experience in new product development and manufacturing engineering, working both in Brazilian industry and in academia. Professor Canciglieri Junior's research has been funded directly by PUCPR, industry and Brazilian funding agencies (*CNPq, CAPES* and *Fundação Araucária* in Brazil). He has strong experience in product design, information modeling to support product design and manufacturing; his work has led more than 160 research publications and some 20 Ph.D. completions. His research is focused on support the design of prosthesis including dental field, design for manufacturing and assembly (DFM/DFA), design for sustainability, concurrent engineering, assistive technology, development of sustainable products focused on the production, generation and use of renewable energy and sustainable development. Professor Osiris Canciglieri Junior is Editorial Member and Reviewer of scientific journals in his research field area.

Abbreviations

ANHD	Human Needs Adequacy Indicator
CAPES	Coordination for the Improvement of Higher Education Personnel
CNPq	National Council for Scientific and Technological Development
DFA	Design for Assembly
DFM	Design for Manufacture
HCD	Human-Centered Design
HNT	Human Needs Theory
HUNE	Human Needs
IPDP	Integrated Product Development Process
ISO	International Organization for Standardization
MOP&RD	Integrated Product Development Model Oriented to Research and Development Projects
PDP	Product Development Process
SI	Social Innovation
UCD	User-Centered Design
UID	User-Interface Design
UX	User Experience

List of Figures

List of Tables

Chapter 1
Introduction

Human beings are the only known creatures on earth that can reason complex things, have intelligence, and create systematically. While they are on earth, they seek to make their life better in some way (Falconi 1999), this includes taking care of themselves, other people, other creatures, and the earth itself.

Throughout human life on earth, constant changes are noticeable, some called evolution and even revolution. The fact is that the human being alone, and especially in the community, is in constant search to improve things, the example of which is that in this process, humanity passed from life in caves to life in cities, with buildings, factories, organizations, in homes with energy, water, sewage, and communication networks. It might not have been the best way to make improvements, but the search continues.

These changes lead to positive and negative consequences and perceptions, and for everything that proves to be negative, in some way, even if only for a portion of society, a contrary movement is generated, the so-called "turning point" (Gladwell 2002). This process is studied in trends and is important for social and human growth. But *"what does this have to do with this research?"*. People are complex and changeable, so it is necessary to understand their needs, in the first place, personalities, values, desires, motivations, behaviors, reactions and emotions to take care of them, supplying their needs and helping to achieve quality of life through solutions suitable.

Considering and involving the people, to whom a design is focused, in the design process, and their relationship and interaction with the solution in each phase of its life cycle, is essential to obtain good results in a design (Vincent et al. 2014), with better quality (Kolus et al. 2018), and better market acceptance, contextual coherence, and economic efficiency during the design process and post-sales creating social value (Choi et al. 2018).

The evolution and continuous creation of innovative technologies are anthropological constants in human social practice (Brödner 2013), new products and services are created with the objective of meeting social needs through innovation (Bennett and McWhorter 2019). These facts, in the current context, may be related to social

G. Unger Unruh and O. Canciglieri Junior, *Human Needs' Analysis and Evaluation Model for Product Development*, https://doi.org/10.1007/978-3-031-12623-9_1

1

innovation (SI) with a focus on improving the quality of people's lives. Although this research does not explore the issue of social innovation, the authors believe that organizations can also see their products or services as social goods and still benefit from economic performance and technological development, contributing to the search for profitable, innovative, and sustainable solutions (Bennett and McWhorter 2019; Balyeijusa 2019).

The Theories of Human Needs (HNT) have proposed explanations regarding human feelings and what motivates human beings in their actions and behaviors in various situations (Milyavskaya and Koestner 2011). These two factors are directly connected to motivation theory, a well-established field of research in psychology that, despite the number of existing studies, is growing in interest because of new discoveries in the area of study, related to the multidisciplinary of these theories being applied, for example, in business practices and everyday experiences, which can add value to research that seeks to understand human aspects in designs (Kispal-Vital 2016). For Smeenk et al. (2019) researches focused on the formalization of methodologies for human needs are expanding, in this context, the present thesis aims to better understand the consideration of the human being in product development, proposing a model of analysis and assessment of human needs in the PDP (Development Process) of Products), to assist in the creation of solutions focused on adapting to people, seeking the essence of caring for social well-being.

In this context, the motivation behind developing products oriented to people's well-being is clear to seek and design solutions to improve their quality of life. Practicing and doing good can be where each person is, in doing an excellent job, doing what you believe in, honestly, aiming for good, being who you are and being your best in this world where beauty is in the diversity of personalities, talents and gifts of each person, that working together can make this world a good place to live, and live what God made each one to live for, a life of love, care, growth and fullness. This means taking care of the well-being of others and the environment, taking care of what is most important in life, relationships, and love. The human being was created naturally creative to make this possible. From this, a hope and a social challenge are born, to which this study wants to contribute by structuring Human-Centered Design (HCD) methods that allow the involvement of people and information about them in the product development process, to create products that their necessities and generate quality of life.

In this way, the theme approached in this book is justified in the theoretical bases of Product Development Process (PDP) and Design Processes focused on technical and structural functions of products, omitting the relationship of products with users, and provides little direction of how and when to deal with user aspects during the process (Coelho 2010). The methodological bases of Usability have been developed providing an increase in the knowledge of the area and its focus on the industry, but there is still a lack of integration of theory in practice (Van Kujik et al. 2007). Although there are several human-focused PDP methods available, and a variety of applications of the concepts, there is still a lack of detailing of design process activities related to the human focus, also in the health area (Harte et al. 2017). The focus on User Oriented Design can be important to achieve strategic objectives of organizations (Veryzer and Mozota 2005), and user involvement in the PDP at the beginning of the design helps to reduce costs in the process and generate successful products (Shluzas and Leifer 2014).

In a study carried out by Nishikawa et al. (2013), it was found that user-generated products systematically and substantially outperformed designer-generated products, of course the process itself was developed by designers and the entire design team, but that means involving users in the PDP, and using methods that involve users in the generation of design solutions for their needs, can bring great advantages in the results of PDPs. On the other hand, Friess (2010) encourages the creation of a model centered on designers rather than on users, due to the intuitions, experiences, capabilities, and specific processes of designers, which are essential for the creation of good solutions. Anay (2011) argues that designers have the ability to translate users' needs into the process of structuring design problems and translate them into qualities of design solutions. This means that neither one nor the other focus must be absolute and exclusive, but both must occur together. In this way, understanding human characteristics, contexts and needs well is essential for the creation of successful products, but you cannot transform people's data into direct solutions, they must be analyzed, interpreted, and translated into solutions through processes suitable, and then validated with the people who may interact with the products.

Flowers (2014) suggested that the purpose of Usability Engineering be changed according to the way of thinking. So, from the point of view of a design, the purpose is to help designers change their perspective, to understand the user's thinking and then design with it in mind. The incorporation of assessment tools in the PDP and the consideration of the human being in its physical and mental aspects becomes a necessary and urgent strategy (Merino et al. 2012), and since the 1980s, methodologies oriented to this end have emerged, but there is still no broad and well-defined view of the human-centered design approach—HCD (Van Eijk et al. 2012).

In the literature, several models, methods, tools, and structures for HCD have already been developed and proposed, some of which are focused on specific types of products and others are a little more comprehensive and detailed, such as the IDEO model (2009) (Van Kijuk 2010; Zeng et al. 2010; Baldassare et al. 2017; Reinert and Gontijo 2017). However, most existing PDPs are rigid, linear, and step-by-step progressive, they do not accommodate existing Usability methods, cannot be adjusted to specific design needs, are limited to applications to specific product types, and do not support generation of necessary documents (Helms et al. 2006).

Models such as the Stage-Gate are useful, but they cannot fully capture the impact of the HCD dynamics on the PDP (Veryzer and Mozota 2005), in the same way that it was perceived in the model proposed by Unruh (2015). Based on this statement, Mao et al. (2005) reinforces the need for a methodology that encompasses the development process from start to finish in a cyclical way.

With this information and with the arguments of Zeng et al. (2010) attesting that creativity and HCD, when worked together, bring benefits to people and businesses. With this, it is possible to perceive the importance of the human-centered approach in a PDP, and to identify the opportunity to systematize methods and structures considering human factors in designs, in a clear and dynamic way. It was clear that there

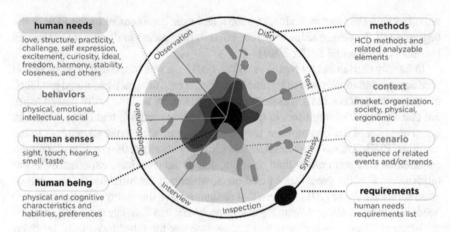

Fig. 1.1 Proposed *HUNE* model. *Source* Unruh (2020)

are several scientific methods available in the literature and that the product development process can become confusing and inefficient, especially when considering its application in the market, because, if it is not well systematized and does not contain a clear form of aid in making decision would be of truly little use. In this way, the research presented in this book sought to identify, evaluate, and organize human needs and aspects, proposing the HUNE Model of Human-Centered Design (HCD) (Fig. 1.1) that was oriented to the Product Development Process (PDP) having based on the theoretical basis of the main concepts related to the topic studied.

References

Anay MÖ (2011) Bridging the gap: designer's "user concept" as a transformative tool between user knowledge and design. Anadolu Univ J Sci Technol 12(2):111–118

Baldassare B, Calabretta G, Bocken N, Jaskiewicz T (2017) Bridging sustainable business model innovation and user-driven innovation: a process for sustainable value proposition design. J Clean Prod 147:175–186

Baleijusa SM (2019) Sustainable development practice: the central role of the human needs language. Soc Change 49(2):293–309

Bennett EE, McWhorter RR (2019) Social movement learning and social innovation: empathy, agency, and the design of solutions to unmet social needs. Adv Dev Hum Resour 21(2):224–249

Brödner P (2013) Reflective design of technology for human needs. AI & Soc, 25th anniversary volume, a faustian Exchange: what is to be human in the era or ubiquitous technology? 28:27–37

Choi Y, Na JH, Walters A, Lam B, Boult J, Jordan PW, Green S (2018) Design for social value: using design to improve the impact of CSR. J Des Res 16(2):155–174

Coelho DA (2010) A method for user centering systematic product development aimed at industrial design students. Des Technol Educ Int J

Falconi V (1999) Controle da qualidade total: no estilo japonês. Editora de Desenvolvimento Gerencial, Belo Horizonte

Flowers J (2014) Usability Engineering can change our thinking. Technol Eng Teach

Friess E (2010) The sword of data: does human-centered design fulfill its rgetorical responsability? Des Issues 26(3)

Gladwell M (2002) O ponto da virada. Little, Brown and Company, USA

Harte R, Glynn L, Rodríguez-Molinero A, Baker PMA, Scharf T, Quinlan LR, Ólaighin G (2017) A human-centered design methodology to enhance the usability, human factors, and user experience of connected health systems: a three-phase methodology. JMIR Hum Factors 4(1)

Helms JW, Arthur JD, Hix D, Hartson HR (2006) A field study of the wheel—a usability engineering process model. J Syst Softw 79:841–858

Ideo (2009) Human centered design toolkit. Available in: https://www.designkit.org/methods. Accessed in: 3rd Mar 2020

Kispal-Vital Z (2016) Comparative analysis of motivation theories. Int J Eng Manag Sci (IJEMS) 1(1)

Kolus A, Wells R, Neumann P (2018) Production quality and human factors engineering: a systematic review and theoretical framework. Appl Ergon 73:55–89

Mao J-Y, Vredenburg K, Smith PW, Carey T (2005) The state of user-centered design practice. Communications of the ACM 48(3)

Merino GSAD, Teixeira CS, Schoenardie RP, Merino EAD, Gontijo LA (2012) Usability in product design—The importance and need for systematic assessment models in product development—USA-Design Model (U-D) ©. Work 41:1045–1052

Milyavskaya M, Koestner R (2011) Psychological needs, motivation, and well-being: A test of self-determination theory across multiple domains. Pers Individ Differ 50(3):387–391

Nishikawa H, Schreier M, Ogawa S (2013) User-generated versus designer-generated products: a performance assessment at Muji. Int J Res Mark 30:160–167

Reinert F, Gontijo LA (2017) Proposta de sistemática para a integração da ergonomia no projeto de produtos. Hum Fact Des 6(12):105–123

Shluzas LMA, Leifer LJ (2014) The insight-value-perception (iVP) model for user-centered design. Technovation 34:649–662

Smeenk W, Sturm J, Eggen B (2019) A comparison of existing frameworks leading to an empathic formation compass for co-design. Int J Des 13(3):53–68

Unruh GU (2015) Método Conceitual de Processo de Desenvolvimento Integrado de Produtos eletrodomésticos orientado para a Usabilidade. Dissertação (Programa de Pós-Graduação em Engenharia de Produção e Sistemas)—Pontifícia Universidade Católica do Paraná, Curitiba

Unruh GU (2020) Modelo de análise e avaliação de necessidades humanas para o desenvolvimento de produtos—PHD Thesis—Systems and Industrial Engineering Postgraduate Program. Pontifical Catholic University of Paraná, Curitiba, Brazil

Van Eijk VD, Van Kujik J, Hoolhorst F, Kim C, Harkema C, Dorrestijn S (2012) Design for usability; Practice-oriented research for user-centered product design. Work 41:1008–1015

Van Kujik J, Christiaans H, Kanis H, Eijk DV (2007) Usability in product development: a conceptual framework. In: Proceedings of the ergonomics society annual conference. Contemporary Ergonomics Nottingham, UK

Van Kujik J (2010) Managing product usability: How companies deal with usability in the development of electronic consumer products. Thesis (Doctorate)—Delft University of Technology. Faculty of Industrial Design Engineering. Netherlands

Veryzer RW, Mozota BB (2005) The impact of user-oriented design on new product development: an examination of fundamental relationships. J Product Innov Manag 22:128–143

Vincent CJ, Li Y, Blandford A (2014) Integration of human factors and ergonomics during medical device design and development: It's all about communication. Appl Ergon 45:413–419

Zeng L, Proctor RW, Salvendy G (2010) Creativity in ergonomic design: a supplemental value-adding source for product and service development. Hum Factors 52(4):503–525

Chapter 2
Human Needs: Analysis and Evaluation Approach for Product Development Context

This chapter has been organized into three sections, which are: (i) The theoretical framework necessary for the design and construction of the HUNE model (Human Needs Theory (HNT), the contextualization of Design, the concepts of Ergonomics, Human Factors, Usability, User-Centered Design (UCD), Human-Centered Design (HCD), User Experience (UX), and the Product Development Process (PDP)); (ii) a reflection on "human being versus human doing"; and (iii) a discussion of what was presented in this chapter from the perspective of the theoretical framework addressed.

2.1 Background

The theoretical background was divided into four parts, the first presents the Theory of Human Needs (HNT), the second a contextualization of Design, the third presents the areas of knowledge that support the study, where the concepts of Ergonomics and Factors are presented (Human Resources, Usability, User-Centered Design (UCD), Human-Centered Design (HCD), User Experience (UX), and Product Development Process (PDP), and the fourth deals with some terms applied in the denomination of people in the product development.

2.1.1 Theory of Human Needs (HNT)

Theories of human needs (HNT's) provide fundamental explanations of human feelings and what motivates individuals' actions and behaviors in various situations (Milyavskaya and Koestner 2011). Alharti et al. (2018) proposed a needs identification model with 12 categories of needs, based on the hierarchy of Ford and Maslow,

---

The original version of this chapter was revised: some of the omitted references were included. The updated version to this chapter is available at https://doi.org/10.1007/978-3-031-12623-9_5

© The Author(s), under exclusive license to Springer Nature Switzerland AG 2023, corrected publication 2023

G. Unger Unruh and O. Canciglieri Junior, *Human Needs' Analysis and Evaluation Model for Product Development*, https://doi.org/10.1007/978-3-031-12623-9_2

they are: structure, practicality, challenge, self-expression, excitement, curiosity, freedom, ideal, harmony, love, closure and stability. Chulef et al. (2001) made a survey of human needs and motivations based on several authors, as shown in Table 2.1, and through a survey with people, they defined categories for these needs and motivations in relation to the number of needs and motivations, listed in the Table 2.2.

People are not entirely rational when it comes to accessing their priorities, the consequences of their actions and what gives them pleasure, as well as they cannot satisfactorily judge their intentions in the quest to understand all this well in order to be able to develop daily solutions (Kispal-Vital 2016). The same author discusses the motivational circle, based on Griffin and Moorhead (2013) as illustrated in Fig. 2.1, where it is possible to observe an endless cycle, arguing that a need leads to the search for methods with the purpose of satisfying these needs, leading to behaviors, which generate rewards and punishments, to a reassessment of necessities, generating need deficiency, because it is not fully satisfied and needing to start the cycle again.

Therefore, and as defended by Bennett and McWhorter (2019), that new products and services are created to meet social needs through innovation, and learning to empathize with the target audience is fundamental for Social Innovation (SI), to evaluate and reassessing people's needs by understanding the methods they use, their behaviors and the reassessments they make, leads to endless opportunities to meet needs, as society constantly changes and their needs will never be fully satisfied, revealing that there will always be the possibility of creating new products and services.

2.1.2 Design

Since the beginning, the design activities aim to create solutions to improve people's lives, that is, it is for the human being, for society, Simon (1969) said that design seeks to change existing situations to preferable ones. Papanek (1971) advocated socially responsible design integrating aspects of anthropology into design. Rittel (1973) defended the importance of human experience and perception in design, presenting phenomenology. And Jones (1978) argued that design is defined from society, the use, and benefits that a product can generate for people, thus, it influences and is influenced by society and all its cultural complexity. Cross (1982) through his mapping of how designers think and make decisions influenced the emergence of the term "design thinking" which led to design being applied in other areas. Sanders (1999) based on his knowledge of psychology and anthropology created several methods and techniques focused on human-centered design and design thinking.

In this social context in which design is inserted, there are several subjects who relate to and influence or are influenced by decisions and design process. Löbach (2001) pointed out four related parties in the design communication process: the designer, the company, the user, and the design object. Bomfim (1995) defined five related subjects: creator (or designer), producer (or organizations that produce the products), consumer (or user of the product), society (social and political institution) and the product. For Baxter (2000) the definitions were: consumers; sellers;

Table 2.1 Human needs derived from psychology literature

Shorthand	Full label
Achieve salvation	Achieve salvation
Art	Appreciate art
Aspirations	Achieve my aspirations
Sexually attractive	Being able to attract, satisfy, sexually arouse a sexual partner
Avoid failure	Avoid failure
Avoid guilt	Avoid feeling guilty
Avoid rejection	Avoid rejection from others
Avoid stress	Avoid stress
Be able to fantasize	Being able to fantasize, imagine
Be affectionate	Be affectionate towards others
Be ambitious	Be ambitious, hard-working
Be better than others	Be better than others, beating others
Be carefree	Be cheerful (light of heart), carefree, enjoying life
Be clean	Be clean, tidy (personal care)
Be conventional	Maintain conventional views, avoid innovation
Be creative	Be creative (e.g., Artistically, scientifically, intellectually)
Be curious	Be curious, inspect, learn
Be disciplined	Being disciplined, able to follow designs I have started, follow my intentions with behavior
Be free	To have freedom (to be a free person)
Be good looking	Be good looking
Be honest	Be honest, faithful, respectful, courteous, considerate of others
Be in love	Be in love
Be innovative	Change my ways, be innovative in the way I live my life
Be smart	Be smart
Be kind	Be kind, make friends, bring others close
Be logical	Be logical, consistent, rational
Be in love	Be really passionate about something
Be playful	Be playful, lively, act for fun
Be popular	Being at the center of things, being popular
Be practical	Be practical
Be private	Keep it to myself, be private
To be recognized	Be admired, recognized by others
Be reflective	Be reflective, not impulsive
Be respected	Be respected by others
Be responsible	Be responsible, trustworthy
Be self-sufficient	Be independent, self-reliant, self-sufficient

(continued)

Table 2.1 (continued)

Shorthand	Full label
Be socially attractive	Be socially attractive, exciting, fascinating, impress others
Be spontaneous	Be spontaneous
To be unique	Be unique, different, exceptional
To belong	Belonging to social groups, feeling part of a group
Accounts	Being able to meet my financial needs, not worrying about bills, expenses, etc.
To buy things	Buy things i want
Be careful	Be careful
Career	Have a career
Career knowledge	Stay up to date with career-related knowledge
Cause	Be committed to a cause (e.g., Planet, environment, anti-crime, anti-drugs)
Charity	Be charitable, help those in need
Nearby children	Being close to my children
Close spouse	Being close to my spouse
Happy with myself	Being happy, content with myself, having inner harmony, freedom from inner conflicts
Contribution	Have an erotic relationship
Environment control	Be an ethical person
Control over others	Have an exciting, stimulating life
Decisions by others	Be physically active, exercise regularly
Defense versus Criticism	Seek new things, explore, be adventurous
Descendants	Follow fashion
Different experiences	Feeling close to my parents, siblings, grandparents
Hard things	Having emotional intimacy, feeling really connected with another person
Easy life	Feel safe
Education	Finding greater meaning in life, coherence, harmony, unity
Educational level	Have firm values
Entertain others	Have flexibility of point of view, open mind and be open
Erotic relationship	Have freedom of choice
Ethic	Have a good marriage
Exciting life	Being a good parent (teaching, providing, passing on values)
Exercise	Having friends, I love, close company
To explore	Receiving help from my parents, siblings, grandparents
Fashion	Helping others, cooperating, supporting
Feel close to my family	Have an erotic relationship
Feel in tune	Be an ethical person
Feel safe	Have an exciting, stimulating life

(continued)

Table 2.1 (continued)

Shorthand	Full label
Find greater meaning	Be physically active, exercise regularly
Firm values	Seek new things, explore, be adventurous
Flexibility	Follow fashion
Freedom of choice	Feeling close to my parents, siblings, grandparents
Good marriage	Having emotional intimacy, feeling really connected with another person
Good father	Feel safe
Have friends	Finding greater meaning in life, coherence, harmony, unity
Family help	Have firm values
Help others	Have flexibility of point of view, open mind and be open
Hobbies	Devoting time to fun, recreation, entertainment, hobbies
Intellectual conversations	Having experiences, intellectual conversations, discussing interesting topics
Job	Having a job, I really like
Meet many others	Knowing and being on familiar terms with many others
Know myself	Know myself, be in touch with myself
Leader	Be a leader
Learn art	Learn more about art
Life limitations	Learning the limitations of life
Live close to family	Living close to my parents, siblings, grandparents
Distinguished opinion	Take good care of me, look distinguished
Look fit	Look physically fit
Look young	Look young
Mature romantic	Having a mature romantic relationship
Mature understanding	Have a mature understanding of life
Mechanical skill	Have mechanical skill
Mental health	Be mentally healthy
Mentor	To have a mentor, someone to guide me
Cash	Make a lot of money
New ideas	Have new, original ideas
Nutrition	Be physiologically healthy, maintain a healthy weight, eat nutritious foods
Others to count	Have others to count on
Trust of others	Have the trust of others
Overcome failures	Overcome failure, get me back on my feet after failure
Own guidelines	Set and follow my own guidelines
Peace of mind	Have peace of mind
Personal growth	Experience personal growth
Persuade others	Influence, persuade others

(continued)

Table 2.1 (continued)

Shorthand	Full label
Physical skill	Have skill, physical agility
Physical aptitude	Being in good physical condition, good physical shape
Please god	Please God
Provide family	Providing my spouse or children (or both) with a sense of financial security and a home to return to
Pursue ideals	Pursue my ideals, fight for the things I believe in
Religious faith	Keep religious faith
Religious traditions	Engaging in religious traditions
Rich social life	Have a rich, active social life
Romantic experiences	Have romantic experiences
Seek equality	Be involved in seeking equality, fraternity, and equal opportunity for all
Seek equity	Seek equity
Seek justice	Seek justice
Self esteem	Having a high self-image, self-esteem, feeling good about myself
Give good examples	Give good examples
Sexual experiences	Enjoy sexual experiences
Share feelings	Share feelings with close friends
Stability	Have a stable life, avoid change, adhere to my ways and lifestyle
Stable family life	Having a stable, secure family life (with my spouse, children, or both)
Defend beliefs	Defend my beliefs
Support from others	Getting support from others on designs I believe in
Take care of my family	Taking care of my parents, siblings, grandparents
Accept risks	Not being fearful, being able to take risks
To teach	Developing others (teaching, sharing knowledge)
Things in order	Keeping things in order (my desk, office, home, etc.)
Think intellectually	Being able to think intellectually (handle data, extract ideas, create hypotheses, analyze, synthesize information)
Well-being	Protect my well-being, avoiding pain
Wisdom	Have wisdom
Beauty of the world	Experiencing a beautiful world (going to museums, concerts, being with nature)

Source Chulef et al. (2001)

Table 2.2 Categorization of human needs and motivations

Category	Numbers of items
Flexibility, openness and excitement	10
Recognition and social approval	8
Personal growth	8
Family	8
Sex and romance	7
Self-sufficiency	7
Stability and security	7
Social consciousness	6
Physical health	5
Physical appearance	5
Leadership	5
Intellect and education	5
Religion	4
Psychological well-being	4
Ethics and idealism	4
Finance	4
Receive from others	4
Teach and help others	4
Aesthetics	3
Career	3
Creativity	3
Positive social qualities	3
Friendship	3
Achievement	3
Training	3
Wedding	2
Defense x rejection	2
Freedom	2
Dexterity	2
Find greater meaning	1

Source Chulef et al. (2001)

production engineers; designers; and entrepreneurs. Melo (2003) defined: client, user, and designer. Each of these subjects has specific characteristics, needs and expectations, which the design process must seek to meet (Baxter 2000; Lidwell et al. 2010). In other words, Human or User-Centered Design (HCD or UCD) has always been intrinsic to design, however, throughout history, some applications and studies aimed at including the human in the design process have been transformed, and it has become necessary to creation of specific disciplines that worked on the

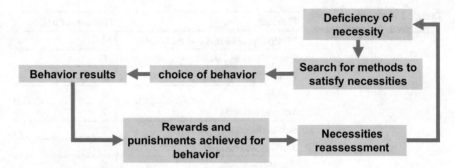

Fig. 2.1 Motivational framework. *Source* Elaborated by the auhors based on Griffin and Moorhead (2013)

human factors, including knowledge from other areas of knowledge. These specific disciplines are presented in the next item.

2.1.3 The Origin of Disciplines

The creation and adaptation of things to human needs and comfort have existed since prehistory, but the discipline that started to formalize and systematize knowledge and processes in this area was Ergonomics or Human Factors, which emerged during the first world war, were improved. In World War II and consolidated in the industrial revolution through its applications in work activities (Iida 2005). Shortly afterwards, the area grew and began to be applied in the most diverse types of products and situations, including digital products and the discipline called Usability has emerged approximately in the mid-nineties (Nielsen 1993).

Along with this evolution, the term User-Centered Design (UCD) was born, focusing on satisfying the fundamental needs of users, covering two dimensions: protection (protecting users from harm); and appreciation. From the 1980s to the 2000s, the concept of protecting the user with the use of Usability Engineering has been diluted into a more versatile range of design objectives, which went from promoting comfort to promoting emotions. This occurred with the emergence of Design for User Experience (UX), where Usability principles are just some of the goals of UCD, which now also cover new experiences, pleasure, and emotion in interacting with products (Keinonen 2010). These disciplines are applied to work processes and environments, physical products, digital products, environments in general, and more recently to services. They are also applied in the areas of marketing and business, the difference being in the moment of human interaction, that is, at the moment of acquiring the product using the term "user" instead of the term "consumer" or "customer".

2.1.4 Ergonomics and Human Factors

According to the Brazilian Ergonomics Society, ergonomics is "the study of the relationship between man and his work, equipment, work environment and particularly, the application of knowledge of anatomy, physiology and psychology in the search for solutions to problems that arise from this relationship". However, for the International Ergonomics Association (IEA 2020), Ergonomics (or human factors) is the scientific discipline concerned with understanding the interactions between human beings and other elements that are part of a system, it applies theories, principles, data and methods aimed at optimizing human well-being and their respective performance in the general system.

Ergonomics and Design had three objectives related to user needs over the years (Hoyos-Ruiz et al. 2015): (i) Provide information to designers throughout the PDP, mainly in the proposal stages, concepts and final evaluations; (ii) Support to understand human beings, their physical and cognitive processes; (iii) Integrate the discipline to engage people throughout the PDP with the intention of understanding people beyond their physical and cognitive processes.

This same society states that the practice of Ergonomics/Human Factors does not belong to a specific domain, it is a multidisciplinary and user-centered integrative science, it takes into account physical, cognitive, socio-technical, organizational, environmental and other relevant factors, as well as the complex interactions between humans and other humans, the environment, tools, products, equipment and technology (IEA 2020). In other words, it is grounded in sociotechnical values, the principles of Ergonomics/Human Factors, which are: (i) human beings as assets; (ii) technology as a tool to help humans; (iii) promotion of quality of life; (iv) respect for individual differences; and (v) responsibility to all interested parties.

This perspective is the most current in Ergonomics because it is based on a holistic view, on the relationship between all aspects related to socio-technical systems, such as safety, health, cognitive and psychosocial aspects of all parts of the system. In addition, it emphasizes the relationship of any person profile in this system, such as influencers (external), system decision makers (who make system decisions), experts (in product areas) and actors (users, workers, employees, etc.).

2.1.5 Usability

Usability emerged from the needs of the evolution of digital technology, when it began to be realized that the people who used the systems were important for the development of interfaces, then the term "user friendly system" emerged. (Nielsen 1993), with the objective of making the systems easier to use. From there, several related terms emerged, such as HCI (Human–Computer Interaction), UCD (User-Centered Design), HMI (Human–Machine Interface), UID (User-Interface Design), among others.

According to ISO 9241-11 (2018), Usability is "the extent to which a product can be used by specific users, to achieve a specific objective with effectiveness, efficiency and satisfaction in a specific context of use". Where:

- *Users* are any people who interact with the product in some way, the context is the situation that involves this interaction, whether physical, cognitive, sensory or otherwise.
- *Efficacy* refers to the completeness of the task, or objective of using the product, accuracy in performing the task, recall of the use of the interface and quality of the result.
- *Efficiency* refers to the resources used to complete the task, which may include: time, input rate (number of data the user enters into the system), mental effort during the interaction, usage patterns (frequency of use of certain parts, information accessed and deviations of steps to reach the optimal solution), communication effort (when there is interaction with other people) and learning (time curve in similar tasks);
- *Satisfaction* refers to user preferences, perception of ease of use, specific attitudes (positive or negative comments) and perceptions of the user himself, other people's attitudes, attitudes related to content, results, and interaction (Hornbaek 2006).

The literature also presents a series of usability principles, criteria and recommendations (Bastien and Scapin 1993; Dul and Weerdmeester 1991; Jordan 1998; Shneiderman 2005; Nielsen 1993) that can be applied to different interfaces, and help specialists to guide your investigations and usability analyses, including aspects such as orientation, driving, workload, user control, error management, adaptability, consistency, among others.

2.1.6 User Centered Design (UCD)

User-Centered Design (UCD) is an umbrella term, where the "centered" part means that aspects of the respective design take place around a centre, the human being; the "design" part refers to the total creation of the human experience, and can also involve "discovering", "defining", "developing", and "delivering" (Veryzer and Mozota 2005). According to ISO 9241:210 (2019), User-Centered Design has two focuses:

(1) *Design (design)*—determine the knowledge, capabilities and limitations of users regarding the tasks for which the product or system is being developed, the particular interest is in understanding the users' tasks and the vocabulary of the tasks, as well as the physical capabilities, among others, and this information is used by designers to maximize usability.

(2) *Evaluate*—access the design in some dimension (interface, functions, recommendations, standards) or compare models (user model, expected time to perform a task, expected usage pattern), with some metrics and data analysis

tools (questionnaires, error log, time log) according to user performance and preferences.

User Centered Design activities include:

- *Make sure User-Centered Design is contained in the design strategy*—represent stakeholders (user), collect market intelligence (information), define and plan system strategy, collect market feedback and analyze trends in users.
- *Plan the Human-Centered Design Process*—consult with stakeholders (users), identify and plan for user involvement, select user-centric methods and techniques, make sure there is a user-centric approach within the development team, plan and manage user-centered design activities, and provide user support.
- *Specify organizational and user requirements*—clarify and document the system's objectives, analyze the users and risks to them, define the use of the system, generate user and organization requirements, and define quality in use objectives.
- *Understand and specify the context of use*—identify and document user tasks, significant user attributes, organizational, technical, and physical environment.
- *Produce design solutions*—assign roles, produce composite task model, explore system design, use existing knowledge to develop design solutions, specify system and use, develop prototypes, and train and support users.
- *Conduct design assessments against requirements*—specify and validate the assessment context, assess initial prototypes to define requirements, assess prototypes to improve the design, assess the system to ensure organizational and user requirements have been met, as well as whether the required practice has been followed, and to ensure that it continues to meet organizational and user needs.
- *Introduce and operate the design*—manage changes, determine impacts on users and the organization, customization, provide training to users, support users in planned activities, and ensure compliance with ergonomic workplace legislation.

2.1.7 Human Centered Design (HCD)

Human-Centered Design is an extension of the UCD, ISO 9241-210 (2019) included "address impacts on a range of stakeholders, not just those normally considered to be users", i.e., it has a broader profile focus of people studied, because it can consider specific or broad profiles, all types of profiles of people who interact with the product and design in some way (stakeholders), directly or indirectly. Even so, in many cases both terms (HCD and UCD) are used synonymously.

In this context, we consider that stakeholders are people or organizations that can affect, be affected by or perceive themselves as affected by a decision or activity, they can include: users, buyers, owners or managers of systems and people who are directly or indirectly affected by the operation of a system, product or service, each of which may have different needs, requirements and expectations (ISO 9241-11 2018). It also stated that HCD is an approach to the design and development of systems that aims to make them more usable, applying knowledge and techniques of ergonomics, human factors, and usability, can bring benefits such as increased productivity, user well-being, accessibility, stress prevention, and harm reduction.

Therefore, it covers aspects of all the areas mentioned above, becoming a broad and inclusive concept, however, from the literature review carried out in this research, it was identified that currently most of the studies found with this focus are aimed at application in medical or social areas, so there is an opportunity to expand its application in other types of organizations that can benefit more people included in their systems.

IDEO (2015) organizes HCD in three phases: (i) Inspiration, learning directly from the people for whom it is designing, deeply understanding their needs; (ii) Ideation, identifying design opportunities and prototyping viable solutions; (iii) Implementation, bringing the solution to life. While ISO 9241-210 (2019) organizes it into four: (i) Identify the user and specify the context of use; (ii) Specify user requirements; (iii) Produce design solutions; (iv) Evaluate design solutions against user requirements. By way of differentiating the terms HCD and UCD, if a product, for example, is developed only with a focus on solving the needs of users (of the people who will use the product after their purchase), this would be a UCD design, now if the same product is developed with a focus on the user and also on other people who will interact with the product (such as salespeople, consumers, people responsible for logistics, maintenance, etc.) this would be an HCD design, and since the international standard defines HCD as an extension of UCD, and in its own definitions describes, HCD could also only consider the user, however it intends to reach more people related to the product, covering its benefits.

2.1.7.1 HCD Methods

There are several HCD methods that can be inserted into a PDP, some of which are presented by classic authors such as Nielsen (1993), Jordan (1998) and others more recently (Maguire 2001; Bevan 2003, 2009; Heinilä et al. 2005; Stanton et al. 2005; Cybis et al. 2007; Roto et al. 2010), among which some even make the results of their survey methods available on websites.

There are also websites developed from studies by some authors, which present several incredibly detailed methods with information on how to apply them, most of them focused on the Usability or UX area, and some even have a system of filters that help in the indication of methods most suitable for application in a case of their need.

Among the websites that present methods and their descriptions, the following were identified: (i) usability.gov, which presents 19 methods, detailed regarding their application and benefits; (ii) design kit from the company Ideo, which presents methods for the entire product development process, already indicating the stage of the process, but there is a mixture of methods, techniques and tips, making the understanding of its application unclear. A great contribution is that each of these items is presented in a detailed way, including the execution steps. Among the methods available on this site, there are also 19 specifically aimed at UCD.

And among the websites that, in addition to presenting the methods, also indicate when to apply them, three more were identified. The first, STRUM—Scheduling Tool for Recommending Usability Methods, is an automated tool still under construction, focused on software development (Cayola and Macias 2018).

The second, UCDToolbox (https://ucdtoolbox.com), is the result of an academic research by Weevers (2012), which presents 34 methods, which can be searched freely or through filters that help in the selection of the most suitable for each design stage and status. In addition to the selection being excellent, all methods have an unbelievably detailed description, facilitating their understanding and application.

The third, Usability Planner (http://www.usabilityplanner.albertoblazquez.net), was also developed by researchers in the area (Ferre and Bevan 2010; Ferre and Bevan 2011), it is a platform for indicating methods for each step of the design, then the site starts with the request to select the step(s) of the design and then presents the list of recommended methods with a brief description and various auxiliary filters to reduce the number of indications. Of all the references found, this one proved to be the most complete and detailed, with 66 methods, it has a filter system that automatically indicates the ideal method for each design situation.

In addition, an article was identified that presents a proposal very similar to the previous ones, but it is a quantitative method for selecting UX techniques (Melo and Jorge 2015), which help in the decision process of which technique to use according to some parameters, including indicated in ISO/TR 16892 (2002), such as cost, time, objectives, type of product, user involvement, number of participants, among others.

The second and third websites mentioned above are ideal for applying HCD methods in product development processes, but both have a greater focus on digital products, and a limited number of methods, among the existing ones. Therefore, nine main authors were identified and analyzed, who present a relevant quantity and quality of methods, to map and categorize the existing methods, in order to facilitate their indication for application in PDPs. The authors selected for this classification were: (1) Nielsen (1993); (2) Jordan (1998); (3) Hom (1998); (4) Maguire (2001); (5) Bevan (2003); (6) Heinilä et al. (2005); (7) Cybis et al. (2007); (8) Bevan (2009); and (9) Roto et al. (2010).

By analyzing these references, it was initially possible to select 89 methods that could be applied to physical, digital products, systems and/or services. As the objective was to generate clarity and assist in the decision process of applying the methods, all of them were listed, analyzed and classified in the following aspects: (i) method category; (ii) results obtained in the method; (iii) authors; and (iv) stage in the product development process. There are many other aspects that influence this decision-making, but these basic ones were enough to help in the direction.

Some authors already make a separation of methods by category, for example, Cybis et al. (2007) categorizes the methods into: (i) Analysis; (ii) Specification; (iii) Generation and organization of ideas; (iv) Design; (v) Evaluation. Analyzing the indications of other authors and the methods themselves, for this study, 10 categories were defined: (i) Inspection; (ii) Diary; (iii) Questionnaire; (iv) Interview; (v)

Observation; (vi) Test; (vii) Synthesis; (viii) Generation of ideas; (ix) Others; and (x) Prototyping. Based on these categories, the methods proposed by the nine selected sources were analyzed and classified, in Table 2.3, into categories of methods and at the time of the PDP in which each method can be applied, as described below:

1. Initial stage of the design, where problems and opportunities related to the target audience profile, context and scenario are researched and identified;
2. Strategy formulation and product briefing;
3. Product development and evaluation during development;
4. Product in use in people's homes.

In addition to this analysis and classification, which already helps in directing the application of the methods, as verified in the references and in ISO/TR 16892 (2002), there are some more variables that can help in the refinement of this direction, they are: cost, objectives, type of product, time, user involvement, number of participants, among others.

The cost is difficult to analyze in the literature, because there are not many reports about the real cost of each method, because it depends on the country, organization, etc., the objectives are linked to the results obtained, which have already been listed in the previous tables, and the product types are both physical and digital, as well as systems, as selected in the reference analysis.

The other variables are shown in Table 2.4. Time was analyzed in the reference data and based on the authors' experiences, user involvement is linked to the number of participants and the place of application, also identified by the same means of the aspect "time". And finally, the indication of the number of evaluators was added, which can be relevant when planning a design.

2.1.8 User Experience (UX)

The basis of User Experience (UX) is multidisciplinary and dynamic, therefore it is used by different areas of knowledge, involves several disciplines, concerns the result of user interaction with the artefact, before, during and after use, and is unique for each person, since it is influenced by previous experiences, in addition to interference from context, internal and external systems, past, present and future (Naumann et al. 2007; Cybis et al. 2007). According to ISO 9241-210 (2019) experiences are "people's perceptions and responses, resulting from the use and/or anticipated use of a product, system or service".

Lund (2006) and Dumas and Salzman (2006) say that the User Experience is an evolution of Usability, according to Padovani et al. (2012) and Beccari and Oliveira (2011) while usability deals specifically with correcting problems, User Experience deals with understanding how people act, think and the reasons for these actions and thoughts. Dumas and Salzman (2006) classify functionality and usability as essential, and pleasure in use as what users expect, which is what User Experience would

Table 2.3 HCD methods

Method	Author									PDP phase			
	1	2	3	4	5	6	7	8	9	1	2	3	4
(1) Inspection (potential problems and opportunities)													
Heuristics/Expert Assessment/Property Checklist/Ergonomics Inspection/Standards	✓	✓	✓	✓	✓	✓	✓	✓		✓			
Diagnostic assessment/Prevention and error inspection/Risk analysis					✓		✓	✓		✓	✓	✓	
Subjective/Analytical/UX assessment		✓			✓		✓			✓	✓	✓	
Functionality		✓	✓	✓				✓		✓	✓	✓	
Hedonic utility scale (HED/UT)								✓	✓	✓	✓	✓	
Perspective based inspection									✓	✓	✓	✓	
Task analysis/Role mapping		✓		✓	✓	✓	✓	✓		✓	✓	✓	
Cognitive step by step		✓	✓		✓	✓	✓			✓	✓	✓	
Macroergonomics (MAS) or Design (MEAD) structure analysis			✓		✓					✓	✓	✓	
Physical ergonomics								✓	✓	✓	✓	✓	
Immersion				✓					✓	✓	✓	✓	
Valuation methods/Cost–benefit		✓						✓		✓	✓	✓	
Requirements comparison								✓		✓	✓	✓	
Analysis of similar or competitors	✓			✓	✓		✓	✓	✓	✓	✓	✓	
(2) Diary (situations, contexts, scenarios, behaviors, understandings)													
Structured									✓	✓	✓	✓	✓
Self-reported									✓	✓	✓	✓	✓
Private camera chat		✓								✓	✓	✓	✓
Narrative audio									✓	✓	✓	✓	✓
Day reconstruction method (DRM)									✓	✓	✓	✓	✓
Context awareness									✓	✓	✓	✓	✓

(continued)

Table 2.3 (continued)

Method	Author									PDP phase			
	1	2	3	4	5	6	7	8	9	1	2	3	4
Affective diary									✓	✓			✓
Incident diary		✓	✓	✓	✓				✓	✓		✓	✓
(3) Questionnaire (opinions, experiences, expectations, and characteristics)													
Profile	✓	✓	✓	✓	✓		✓		✓	✓			✓
Semantic differential									✓	✓		✓	✓
Product semantic analysis (PSA)									✓	✓		✓	✓
AttrakDiff									✓	✓		✓	✓
Intrinsic motivation inventory (IMI)									✓	✓		✓	✓
Experience sampling method (ESM)									✓	✓		✓	✓
Geneva appraisal questionnaire									✓	✓		✓	✓
QSA GQM/SUMI									✓	✓		✓	✓
Differential of emotions (DES)									✓	✓		✓	✓
Perceived comfort rating									✓	✓		✓	✓
Cognitive workload/mental effort				✓				✓		✓		✓	✓
Completing the sentence									✓	✓		✓	✓
Aesthetic scale									✓	✓		✓	✓
Affective grid									✓	✓		✓	✓
Product personality attribution									✓	✓		✓	✓
2DES				✓	✓					✓		✓	✓
Geneva emotion wheels							✓			✓		✓	✓
Satisfaction	✓	✓	✓	✓	✓				✓	✓		✓	✓
Checklists or Emotions cards/Emocards/Emofaces/Emoscope													

(continued)

Table 2.3 (continued)

Method	Author									PDP phase			
	1	2	3	4	5	6	7	8	9	1	2	3	4
SE Expressing experiences or emotions									✓	✓		✓	✓
(4) Interview (opinions, experiences, expectations, and characteristics)													
Traditional	✓		✓	✓	✓	✓	✓			✓		✓	✓
Contextual/Experience investigation/Exploratory testing		✓	✓	✓	✓	✓				✓		✓	✓
Focus group	✓		✓	✓	✓	✓	✓			✓		✓	✓
Post Experience/User feedback	✓							✓		✓		✓	✓
UX scale									✓	✓		✓	✓
This or that									✓	✓		✓	✓
Early experience Assessment (AXE)									✓	✓		✓	✓
Pluralistic step by step								✓	✓	✓		✓	
I.D									✓	✓		✓	✓
(5) Observation (situations, contexts, scenarios, behaviors, understandings)													
Controlled/Assisted evaluation				✓						✓		✓	
Field/Observation/Ethnography	✓		✓	✓		✓	✓			✓		✓	✓
Live laboratory	✓						✓	✓		✓		✓	✓
(6) Test (situations, contexts, scenarios, behaviors, understandings)													
Traditional/Laboratory/Controlled	✓	✓	✓	✓		✓				✓		✓	✓
Co-discovery		✓	✓							✓		✓	
Field	✓	✓	✓	✓		✓	✓			✓		✓	✓
Experience clip	✓								✓	✓		✓	✓
(7) Synthesis (synthesis of research data)													

(continued)

Table 2.3 (continued)

Method	Author									PDP phase			
	1	2	3	4	5	6	7	8	9	1	2	3	4
Storyboarding/Graphic Narrative				✓	✓	✓	✓	✓		✓		✓	
Mind map/Repertoire grid									✓	✓		✓	
Contexts of use				✓						✓		✓	
Usage scenarios				✓	✓	✓	✓	✓		✓		✓	
User profiles/Personas				✓			✓	✓		✓		✓	
Stakeholder identification or consultation				✓						✓		✓	
User journey										✓		✓	
Guidelines/Design/Style standards				✓	✓	✓		✓		✓		✓	
(8) Generation of ideas (ideas)													
Brainstorming (brainstorming)			✓	✓								✓	
Card sorting			✓	✓	✓	✓						✓	
Future workshop							✓	✓				✓	
Parallel design	✓			✓	✓			✓				✓	
Affinity diagram				✓	✓				✓			✓	
RGT									✓			✓	
Participatory workshops	✓	✓		✓				✓				✓	
(9) Others (experiences, reactions, and emotions)													
Thinking out loud protocols										✓		✓	✓
Metrics or performance model					✓			✓		✓		✓	✓
Eye tracking								✓	✓	✓		✓	✓
Physiological arousal via electrodermal activity									✓	✓		✓	✓
PAD									✓	✓		✓	✓

(continued)

Table 2.3 (continued)

Method	Author									PDP phase			
	1	2	3	4	5	6	7	8	9	1	2	3	4
EMO2									✓	✓		✓	✓
Psychophysiological measurement									✓	✓		✓	✓
Emotion sampling device (ESD)									✓	✓		✓	✓
Valence method									✓	✓		✓	✓
Positive and negative affection variation									✓	✓		✓	✓
Scale/UX curve									✓	✓		✓	✓
Pregnancy scale									✓	✓		✓	✓
SAM									✓	✓		✓	✓
AWARD									✓	✓		✓	✓
Sexy assessment tool (shape)									✓	✓		✓	
(10) Prototyping/Simulation													

Table 2.4 Resources for the types of methods

Method type	Participants	Evaluators	Time	Environment
Inspection	0	2+	1–10 days	Office
Daily	5–10	2–3	3–24 weeks	Field
Quiz	50+	1–3	1–4 weeks	Office
Interview	10+	1–3	1–12 weeks	Field/Laboratory/Meeting room
Observation	5–10	2–3	1–24 weeks	Field/Laboratory
Test	5–10	2–3	1–12 weeks	Field/Laboratory
Synthesis	0	1+	1–12 weeks	Office
Generation of ideas	0+	2+	1–24 weeks	Office

involve. Teague and Whitney (2002) and Hancook et al. (2005) complement, saying that the User Experience encompasses aspects of usability, aesthetics, sensations, emotions, and motivations.

2.1.9 Product Development Process (PDP)

The PDP is the systematization of an activity necessary for the industry to identify the market, user needs and analyze technological constraints, considering business strategies to develop products that satisfy these needs and are compatible with manufacturing (Pugh 1991). For Löbach (2001) the PDP is the development and implementation of an idea to solve problems arising from human needs. Therefore, the PDP is based on people's needs and on the analysis of their interaction cycle with the product. This process is complex and dynamic, requires interactions from different areas and activities of the company, and has a large amount of information (Pereira 2014), and for this reason the systematization of a PDP structure can provide ease in interactions and in the establishment and organization the actions and activities necessary to fulfill the design objective (Pereira and Junior 2014; Pereira et al. 2014).

The literature presents several approaches to PDP, and El Marghani (2011) presents some of these approaches (Table 2.5): Concurrent Engineering, Concurrent Engineering, Stage-Gate, Integrated Product Development Process (IPDP) and Product-Based Business (PBB). As a PDP requires a holistic view of the product, involving multidisciplinary teams in the design, such as marketing, engineering, design and research and development (R&D), considering competition, new technologies, quality demand in products by customers, in which the customer, manufacturer and supplier directly influence definitions, design, evaluations, tests and manufacturing in an integrated manner (Fernandes 2013), the approach used in this research will be the Integrated Product Development Process (IPDP).

The initial activities of an IPDP require robustness and quality in design information, user and market analysis and evaluation, since, according to Rozenfeld et al.

Table 2.5 Approaches about IPDP and their related authors

Approach	Characteristics/Features	Authors
Simultaneous engineering (Concurrent engineering)	Concatenation of interdependent steps, simultaneity between them and adaptable process control tools as needed	Clark and Fujimoto (1991) Miller (1993) Prasad (1996) Hubka and Eder (1988) Pahl and Beitz (1996)
Stage-Gate	It stands out for presenting the concept of control tests (Gates) associated with the stages of development (Stages)	Cooper (1993) Cooper (2001) Clark and Wheelwright (1992) Clausing (1993)
Integrates product development process (IPDP)	It expands the concepts of concurrent engineering to all areas involved in product development, not just engineering functions	Andreasen and Hein (1987) Prasad (1997) Pugh (1990) El Marghani (2011)
Product bases business	Relates the product life cycle to the innovation process	Roozenburg and Eekels (1995) Patterson and Fenoglio (1999) Crawford and Benedetto (2010) Baxter (2000) Rozenfeld et al. (2006)

Source El Marghani (2011)

(2006) the initial phases of PDP are the most expensive (85% of the final cost) and offer greater risk and uncertainty. The same author argues that this type of information should be captured and compiled continuously throughout the product use cycle, in order to feed future designs or improvements in the current design. Therefore, the constant updating of the needs and behavior of the market, through the quality of the information, and the verification of the adequacy of the products to the users, the market and the company's strategy can make a substantial difference in the costs and quality of a design.

Pereira (2014) compiled the main PDP models in a table to analyze their similarities and limitations to develop a new model. The models emphasize the systematization of the process by phases that start with the identification of a need or the generation of an idea, go through the detailing of the proposal, development, production, and end with the launch, monitoring, and discontinuation of the product.

Many of these phases are repeated, but with different terminologies, revealing the difference in the authors' points of view, and from the analysis of these models, the author identified 14 stages of activities grouped into six phases of development, contained in three macro phases, and based on them, he created the Integrated Product Development Model Oriented to R&D Projects in the Brazilian Electric Sector—MOP&D (Table 2.6).

Table 2.6 Phases and macrophases of Pereira's MOP&D Model (2014)

Initiation (startup)	Planning		Design					Deployment			Distribution	Maintenance	
1	2	3	4	5	6	7	8	9	10	11	12	13	14
Statement of demand	Scope definition	Design Planning	Study of principles	Conceptual Design	Preliminary Design	Detailed Design	Design Refinement	Manufacturing process design	Product manufacturing and finishing	Marketing plan	Product launch	Post-launch reviews	Discontinue the product/reverse engineering

2.2 Being Human Versus Doing Human

When researching the Human-Centered Design (HCD) area, it is possible to verify that different authors in different disciplines give specific names to the people for whom a product is developed, the main ones are target audience, client, consumer, user, and stakeholder. In product development, each of these names is given according to the focus of the design, stage, or study, with each discipline and/or area of knowledge using a specific name. It is clear that each discipline needs to have a focus, as well as each stage and component of a design, in order to be effective. But care must be taken in its application because words can have a profound influence on social, organizational and design practice.

The words used to describe who interacts with the products or services of an organization are metaphors that indicate how people are conceived by that organization (McDonald 2006), these words said in practice weigh more than dictionary concepts and are important in social practice of a work in social relations (McLaughlin 2009).

Product development, especially when focused on people, must have a multidisciplinary approach (Vredenburg et al. 2002), so each area involved in a design needs to be clear about its role, and it is important to define the meaning of each term used to name the people for whom the design is developed, as well as its application.

Table 2.7 shows the definitions identified in the Cambridge dictionary and in international standards (ISO 9241-11 2018; ISO/IEC 15288:2008 2008), and the applications based on the analysis of the results of the bibliographic research carried out by Unruh and Canciglieri Jr (2018), considering the terms that appeared the most in their research: target audience, customer, consumer, user and stakeholder. The areas of knowledge were considered based on the identification of the principal areas that make up multidisciplinary teams in product development (Ulrich and Eppinger 2004; El Marghani 2011): marketing, design, engineering, operations, logistics, after-sales.

The consumer is the one who researches and interacts with sales communications, purchases a product and consumes it; the customer is basically the same as a consumer, but also a user, because he is the one who interacts with some service; the user is the one who uses the product, performs or requires maintenance and disposes of it; and stakeholder is the person interested in or who exerts influence on the design or product in any scope, so it involves the people who develop the product, suppliers, people who work in the production, sale and maintenance, user, people who make evaluations and communication of the product, among others. These names identify people by what they do, by their relationship with the design or business, reflecting the current social model and behavior of consumption. They are descriptive of a specific relationship (McLauglin 2009). Figure 2.2 illustrates the different moments of interaction with a product, based on the product cycle (Guimarães 2012) and on the interaction cycle (Van Kujik 2010), where it is possible to visualize these terms in relation to the respective activities.

Not necessarily for the same product, these roles will be played by the same person, nor necessarily by different people. For example, the person who buys the

Table 2.7 Definitions of terms for people

Term	Definition	Knowledge area	When it is applied
Target public or target audience	The specific group of people to whom an advertisement, product, website, or television or radio program is targeted	Marketing Design Coordination	Business plan Business and design strategy briefing Product development Communication
Customer	The person who receives a service	Marketing Logistics After-sales	Business plan Business and sales strategy
Consumer	The person who buys goods or services	Marketing Logistics	Business plan Sales strategy and brand and image positioning Product development Distribution
User	The person using the product, machine, or service Person who interacts with the product	Design Engineering After-sales	Business and design strategy Business plan Product development Maintenance Discard
Stakeholder	Person or organization that has any right, participation, claim or interest in a system or in features that meet its needs and expectations	Marketing Design Engineering Operations Logistics After-sales	Business plan Business strategy, design, sales and brand and image positioning Product development Communication Distribution Maintenance Discard

product is not necessarily the one who uses it, but they can be. In this reality, there are different profiles of people who interact with the products at different intensities and frequency, and they should also be considered in the design. The use of these terms comes from both the consumerist tradition of the 1990s and the democratic tradition that developed people's participation in society and to ensure the adequacy of services (McLaughlin 2009), that is, it is a way of transmitting the concept of new roles. Social conditions that emerged with capitalism. Socially, the use of these terms privileges an aspect of human identity, denying the multiplicity of identity and relationships, socially reflects on the feeling of belonging—a basic human need (Maslow 1943), through actions of purchase and use. One must consider whether there are no better ways of applying names to people that reflect their nature and essence (McLaughlin 2009).

An approach focused on consumption only runs the risk of trying to create needs that do not always exist, taking advantage of social and behavioral information (why people buy things) of human vulnerabilities (acceptance, belonging, frustrations, among others), to suggest something and sell things that may not be good for people,

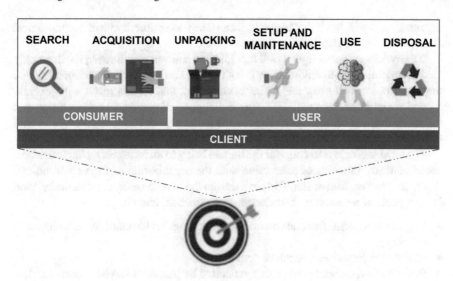

Fig. 2.2 Human interactions and terms

society and the environment, with a one-sided objective of sustaining capitalism, and generating profit (Marx 2011).

For an analysis of the relationship system of these different focuses, observe Fig. 2.3, if the design aimed at the human being as a consumer is very good, but for the moment when a human being will use the product it is not so good, the product and the organization will generate a bad experience for this human, who may, to the detriment of that, decide to discard the product and/or never buy something from the same organization again (Marcus 2004).

If the design for the consumer is very bad, but for the user it is good, the product may take time to make a success in the market, if it does. If you do, it can generate a good experience for your users and they may want to buy products from the same

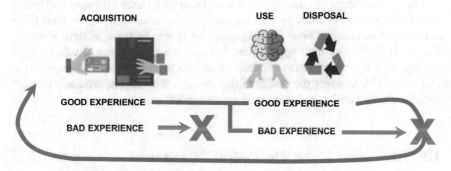

Fig. 2.3 Relationship between experiences and moments of interaction

organization again, but if that happens, it can take a long time, and with technological changes this can be decisively a failure.

If the product is not designed well for either the consumer or the user, it will quickly lead the organization to failure. Now if both focuses do very well, the probability that the organization will grow and be successful, and, that all this result will generate a good social impact is great. Investing in the good of people, in collaboration and sustainability, can bring benefits and returns to the organizations themselves as well as to society (Roca-Puig 2019).

The ideal scenario is to consider the human being in all its aspects, characteristics, social context, moments of interaction with the organization and the environment. According to Pontiano et al. (2014) in a design the human being must be understood in its aspects of personality, knowledge, and expertise, including:

- Cognitive system (mental processes including understanding, learning and memory);
- Motivation (impulses to achieve something);
- Preferences (personal preferences generated by past or observed experiences);
- Social behavior (behavior and activity generated by community life).
- Emotion (perception of desirable events, concerns mood, affection, feeling and opinion);
- Individual differences (differences between people in physical aspects, knowledge, abilities, and skills);
- Intra-individual variability (short-term fluctuations resulting from oscillation, inconsistency, and noise);
- Inter-individual changes (lasting personal changes resulting from learning, development, or aging).

The use of terms related to what people do is useful in defining the focus of efforts in each area and stage of development of products and organizations, it facilitates the organization of tasks and activities of each discipline, as well as of the areas of knowledge, therefore in this meaning the terms "target audience", "customer", "consumer" and "user" should continue to be used for organizational definitions.

But it must always be remembered that all these terms refer to people, and therefore, one must always consider their human characteristics, focusing on their good in the broad spectrum of time, thinking about the future, for these reasons, although this thesis focuses on their applications in users, the term "Human" will be used because it considers the broad spectrum of human beings and it is believed that it will be possible to apply the results of this research with anyone who will interact with the product during its life cycle in future applications.

2.3   Discussion About Theoretical Background

The theoretical framework obtained in the systematic review of the literature was fundamental in conducting the entire research and analysis, and clearly, the fact that it was systematic, had exploratory scientific objectives, definition of terms, criteria, and registration of the process, brought a robust foundation to the study composed of references relevant to the topic addressed.

The process throughout the developed research had several stages, initially influenced by an initial problem and previous objective, to create a new PDP model, which also influenced the analysis of data and references, but after its application, it led the researchers to a different, more open perspective that it was possible to understand the real limitations and opportunities existing in the study area. This process showed that care must be taken with the perspective from which the research is applied, so that it is not influenced in any way, which may lead to an incoherent result.

And this process was also positive, because it made it possible to reanalyze the data, changing the perspective and identifying with greater clarity the real opportunity, instead of a PDP model, a model that helps existing PDPs in the analysis and evaluation with a focus on human needs. And this was only possible because the systematic review of the literature was robust and reapplied for constant updates throughout the development of the research, that is, because a preliminary model was developed and applied, enabling its analysis that led to the proposal of a new refined model. More objective, which was also applied and analyzed in a critical way, and which, after verifying that some aspects were still not consistent with respect to existing studies, led to a reanalysis of all references, applications, processes, and evaluations, in a judicious and changing way the perspective, which resulted in a new model proposal, finally coherent, clear, robust and with a more appropriate and modern look.

Another important factor in the proposal was the process of creating and evaluating human needs requirements, which contributes to the decision making regarding the suitability of products to the human being in a clearer and more systematic way than the methods, models and structures existing until now time. In summary, meeting human needs becomes a basic principle for product development, and human-centered design (HCD) includes several specific areas of knowledge already well consolidated in theories, studies, applications, and standards to assist in this process, and which may be applicable to the development of products and services. Its application requires clarity and systematization so that the process is increasingly efficient, generating positive and promising results.

References

Alharthi R, Guthier B, El Sadik A (2018) Recognizing human needs during critical events using machine learning powered psychology-based framework. IEEE Access 6:58737–58753

Andreasen MM, Hein L (1987) Integrated product development. 2nd edn. The Institute for Product Development, Technical University of Denmark, Lyngby

Bastien C, Scapin DL (1993) Ergonomic Criteria for the evaluation of human computer interfaces. RT-0156 INRIA—Institut National de Recherche em informatique et em Automatique 79

Baxter M (2000) Projeto de produto: Guia prático para o design de novos produtos. Editora Edgard Blücher, São Paulo

Beccari MN, Oliveira TLA (2011) A philosophical approach about user experience methodology. Design, user experience and usability. Theory, methods, tools, and practice. In: Proceeding, Part I: first international conference, DUXU 2011. Orlando, FL, USA

Bennett EE, McWhorter RR (2019) Social movement learning and social innovation: empathy, agency, and the design of solutions to unmet social needs. Adv Dev Hum Resour 21(2):224–249

Bevan N (2003) UsabilityNet methods for user centered design. Human-computer interaction: theory and practice. Proc HCI Int 1(1):434–438

Bevan N (2009) Criteria for selecting methods in user-centered design. Workshop paper I-USED

Bomfim GA (1995) Metodologia para desenvolvimento de projetos. Editora Universitária, João Pessoa

Cayola L, Macías JA (2018) Systematic guidance on usability methods in user-centered software development. Inf Softw Technol 97:163–175

Chulef AS, Read SJ, Walsh DA (2001) A hierarchical taxonomy of human goals. Motiv Emot 25(3):191–232

Clark KB, Fujimoto T (1991) Product development performance: strategy, organization and management in the world auto industry. Harvard Business School Press, Boston, 400 p

Clark K, Wheelwright SC (1992) Revolutionizing product development: quantum leaps in speed, efficiency, and quality. The Free Press. New York

Clausing D (1993) Total quality development: a step-by-step guide to world-class concurrent engineering. ASME, New York

Cooper RG (1993) Winning at new products: accelerating the process from idea to launch. Perseus Books, Reading

Cooper RG (2001) Winning at new products: accelerating the process from idea to launch, 3rd edn. Perseus Publishing, Reading

Crawford M, Di Benedetto A (2010) New products management, 10th edn. McGraw-Hill, Irwin

Cross N (1982) Designerly ways of knowing. Des Stud 3(4):221–227

Cybis W, Betiol AH, Faust R (2007) Ergonomia e Usabilidade: conhecimentos, métodos e aplicações. Novatec Editora, São Paulo

Dul J, Weerdmeester B (1991) Ergonomia Prática. Edgar Blücher, São Paulo

Dumas JS, Salzman MC (2006) Usability assessment methods. Rev Hum Factors Ergon 2

El Marghani VGR (2011) Modelo de Processo de design. Blücher Acadêmico, São Paulo

Fernandes PT (2013) Método de desenvolvimento integrado de produto orientado para a sustentabilidade. Dissertação de mestrado (Programa de Pós-Graduação em Engenharia de Produção e Sistemas)—Pontifícia Universidade Católica do Paraná, Curitiba

Ferre X, Bevan N (2011) Usability planner: Development of a tool to support the process of selecting usability methods. Proceedings of Interact

Ferre X, Bevan N, Escobar TA (2010) UCD method selection with usability planner. In: Proceedings NordiCHI, Reykjavik, Iceland

Griffin RW, Moorhead G (2013) Organizational behavior. Thomson South-Western, Mason, USA

Guimarães LBM (2012) Sociotechnical design for a sustainable world. Theor Issues Ergon Sci 13(2):240–269

Hancock P, Pepe A, Murphy L (2005) Hedonomics: the power of positive and pleasurable ergonomics. Ergonomics Des 13(1)

Heinilä J, Strömberg H, Leikas J, Ikonen V, Iivari N, Jokela T, Aikio KP, Jounila I, Hoonhout J, Leurs N (2005) User centered design: guidelines for methods and tools. The Nomadic Media Consortium

Hom J (1998) The usability methods toolbox handbook, 1998. Available in: http://jthom.best.vwh.net/usability/usable.htm. Accessed in: 22nd Sept 2018

Hornbaek K (2006) Current practice in measuring usability: challenges to usability studies and research. Int J Hum-Comput Stud. 64(2):79–102. https://doi.org/10.1016/j.ijhcs.2005.06.002

Hoyos-Ruiz J, Martínez-Cadavid JF, Osorio-Gómez G, Mejía-Gutiérrez R (2015) Implementation of ergonomic aspects throughout the engineering design process: human-artefact-context analysis. Int J Interact Des Manuf 11:263–277

Hubka V, Eder WE (1988) Theory of technical systems. Springer, Berlin

IDEO (2015) The field guide to human-centered design: design kit. Estados Unidos: IDEO. Disponível em: https://www.designkit.org/methods. Accessed 15 Mar 2022

IEA (2020) What is ergonomics. Available in: https://iea.cc/what-is-ergonomics/. Accessed in: 8th Sept 2020

Iida I (2005) Ergonomia: Projeto e Produção. 2nd ed. Edgard Blücher, São Paulo

ISO 9241-11:2018 (2018) Ergonomics of human-system interaction. Part 11: usability: definitions and concepts

ISO 9241-210:2019 (2019) Ergonomics of human-system interaction. Part 210: human-centred design for interactive systems

ISO/IEC 15288:2008 (2008) Systems and software engineering—system life cycle processes

ISO/TR 16892:2002 (2002) Ergonomics of human-system interaction—Usability methods supporting human-centered design

Jones JC (1978) Métodos de diseño. Editorial Gustavo Gili, Barcelona

Jordan PW (1998) An introduction to usability. Taylor & Francis Ltda, London

Keinonen T (2010) Protect and appreciate—Notes on the justification of user-centered design. Int J Des 4(1)

Kispal-Vital Z (2016) Comparative analysis of motivation theories. Int J Eng Manag Sci (IJEMS) 1(1)

Lidwell W, Butler J, Holden K (2010) Universal principles of design: 125 ways to enhance usability, influence perception, increase appeal, make better design decisions, and teach through design, revised and updated. Rockport Publishers, Beverly, 272 p

Löbach B (2001) Desenho Industrial: base para configuração dos produtos industriais. 1. ed. Edgar Blücher, São Paulo

Lund AM (2006) Post-modern usability. J Usability Stud 2(1)

Maguire M (2001) Methods to support human-centered design. Int J Hum Comput Stud 55:587–634

Marcus A (2004) The ROI of usability, cost-justifying usability, 2nd edn. Elsevier, North Holland

Marx K (2011) O capital: crítica da economia política. Civilização Brasileira, Rio de Janeiro

Maslow AH (1943) A theory of human motivation. Psychol Rev 50:370–396

McDonald C (2006) Challenging social work: the context of practice. Palgrave Macmillan, Basingstoke

McLaughlin H (2009) What's in a name: 'Client', 'Patient', 'Customer', 'Consumer', 'Expert by Experience', 'Service User'—What's Next? Br J Soc Work 39(6):1101–1117

Melo FH (2003) O processo do projeto. O valor do design: guia ADG Brasil de prática profissional do designer gráfico. Editora SENAC, São Paulo

Melo P, Jorge L (2015) Quantitative support for UX methods identification: how can multiple criteria decision-making help? Univ Access Inf Soc 14:215–229

Milyavskaya M, Koestner R (2011) Psychological needs, motivation, and well-being: a test of self-determination theory across multiple domains. Pers Individ Differ 50(3):387–391

Miller LC (1993) Concurrent engineering design: integrating the best practices for process improvement

Naumann A, Hurtienne J, Israel JH, Mohs C, Kindsmüller MC, Meyer HA, Hußlein S (2007) Intuitive use of user interfaces: defining a vague concept. Engineering psychology and cognitive ergonomics. Springer, Heidelberg

Nielsen J (1993) Usability engineering. Morgan Kaufmann, San Diego

Padovani S, Schlemmer A, Scariot CA (2012) Usabilidade & User experience, usabilidade versus user experience, usabilidade em user experience? Uma discussão teórico-metodológica sobre comunalidades e diferenças. Anais do 12º Ergodesign USIHC, 12 a 16 de agosto, Natal-RN, Brasil

Pahl G, Beitz W (1996) Conceptual design. In: Wallace K (eds) Engineering design. Springer, London. https://doi.org/10.1007/978-1-4471-3581-4_6

Papanek VJ (1971) Design for the real world: human ecology and social change. Random House, USA

Patterson ML, Fenoglio JA (1999) Leading product innovation: accelerating growth in a product-based business. Wiley, New York

Pereira JÁ (2014) Modelo de Desenvolvimento Integrado de Produto orientado para projeto de P&D do setor elétrico brasileiro. Tese (Programa de Pós-Graduação em Engenharia de Produção e Sistemas)—Pontifícia Universidade Católica do Paraná, Curitiba

Pereira JA, Junior OC (2014) Product development model oriented for the R&D projects of the brazilian electricity sector. Appl Mech Mater 518:366–373

Pereira JA, Canciglieri Junior O, Lazzaretti AE, Souza MP (2014) Application of integrated product development model oriented to R&D projects of the brazilian electricity sector. Adv Mat Res 945–949:401–409

Prasad B (1996) Concurrent engineering fundamentals, vol 1: integrated product and process organization. Concurrent Engineering. Prentice-Hall PTR, International Series in Industrial and Systems Engineering

Prasad B (1997) Product development methodology. In: Concurrent Engineering Fundamentals, vol II. Prentice Hall PTR, 490 p. 10.13140/RG.2.1.3616.9840

Ponciano L, Brasileiro F, Andrade N, Sampaio L (2014) Considering human aspects on strategies for designing and managing distributed human computation. J Internet Serv Appl 5(10)

Pugh S (1990) Total design—integrated methods for successful product engineering, 1st edn. Prentice Hall, 296 p

Pugh S (1991) Total Design: integrated methods for successful product engineering. Addison Wesley, Massachusetts

Rittel H (1973) The state of the art in design methods: design research and methods. Des Methods Theor 7(2):143–147

Roca-Puig N (2019) The circular path of social sustainability: an empirical analysis. J Clean Prod 212:916–924

Roto V, Lee M, Pihakala K, Castro B, Vermeerem A, Law E, Väänänen-Vainio-Mattila K, Hoonhout J, Obrist M (2010) All UX evaluation methods. All about UX. Accessed 22 Nov 2018 on http://www.allaboutux.org/

Rozenfeld H, Silva SL, Scalice RK, Forcellini FA, Alliprandini DH, Amaral DC, Toledo JC (2006) Gestão de desenvolvimento de produto: uma referência para a melhoria do processo. Editora Saraiva, São Paulo

Roozenburg NFM, Eekels J (1995) Product design: fundamental and methods, 2nd edn. Wiley, Chichester

Sanders E (1999) Design for experiencing: new tools. Proceedings Design and Emotion

Shneiderman B (2005) Designing the user interface: strategies for effective human-computer interaction. 4th ed. Addison Wesley

Simon HA (1969) The sciences of the artificial. MIT Press, Cambridge

Stanton NA, Hedge A, Brookhuis K, Salas E, Hendrick HW (2005) Handbook of human factors and ergonomics methods. CRC Press, Boca Raton

Teague RC, Whitney HX (2002) What's love got to do with it: Why emotions and aspirations matter in person-centreddesign. User Experience 1(3):6–13

Ulrich KT, Eppinger SD (2004) Product design and development. McGraw-Hill, New York

Unruh GU, Canciglieri Junior O (2018) Human and user-centered design product development: A literature review and reflections. Adv Transdisciplinary Eng IOS Press 7:211–220

Van Kujik J (2010) Managing product usability: how companies deal with usability in the development of electronic consumer products. Thesis (Doctorate)—Delft University of Technology, Faculty of Industrial Design Engineering. Netherlands

Veryzer RW, Mozota BB (2005) The impact of user-oriented design on new product development: an examination of fundamental relationships. J Product Innov Manag 22:128–143

Vredenburg K, Isensee S, Righi C (2002) User-centered design: an integrated approach. Prentice-Hall, New Jersey

Weevers T (2012) Web application for UCD Method Selection. Available in: http://ucdtoolbox.com/browse-methods/. Accessed in: 22nd Sept 2018

Chapter 3
Human Needs Model (Hune)

Based on the analysis of methods and models of Human-Centered Design (HCD) existing in the literature, it was possible to notice some similarities and relevant aspects for a design that they bring in familiar, as illustrated in Fig. 1.1. First, they must be defined design requirements according to the needs and opportunities of users and their contexts, by collecting data on their characteristics and behaviors (ISO 9241-210 2019; Nielsen 1993; Jordan 1998; Maguire 2001; Bevan 2003, 2009; Heinilä et al. 2005; Van Kujik 2010; Coelho 2010; Zeng et al. 2010; Gherardini et al. 2016).

Regarding human characteristics, Zeng et al. (2010) (Fig. 3.1—Detail "A") Among several aspects that present, those that most consider human beings are the ergonomic aspects of design: functionality, safety, usability and affectivity; Harte et al. (2017) (Fig. 3.1—Detail "B") and ISO 9241-210 (2019) (Fig. 3.1—Detail "C") include user requirements as a starting point in the Product Development Process (PDP). Van Kujik (2010) (Fig. 3.1—Detail "D") presents the moments of user interaction with the product: exposure (advertisement, word of mouth, approximation); acquisition (observation and exploration); configuration (unpack and install); use (interaction, exposure, and system); maintenance (service and repair); end of life (abandonment and disposal). Furthermore, it indicates aspects of human interaction that must be considered in the PDP: product (its properties, functions and services offered through the product); user (characteristics and capabilities of the user group, physical and cognitive skills, objectives and expectations, diverse or focused group); context (in which physical and social context the product will be used); product use (what are the distinctions of the product's use phases and what usability problems can be expected in each phase). Gherardini et al. (2016) (Fig. 3.1—Detail "E") presents, at various points in its process, the verification of the consistency of the results with the users' needs.

Coelho (2010) presents the implications of use defended by Karlsson and Åhlström (1996): (i) Objective (use for what?); (ii) instrument (use what?); (iii)

The original version of this chapter was revised: some of the omitted references were included. The correction to this chapter is available at https://doi.org/10.1007/978-3-031-12623-9_5

© The Author(s), under exclusive license to Springer Nature Switzerland AG 2023, corrected publication 2023
G. Unger Unruh and O. Canciglieri Junior, *Human Needs' Analysis and Evaluation Model for Product Development*, https://doi.org/10.1007/978-3-031-12623-9_3

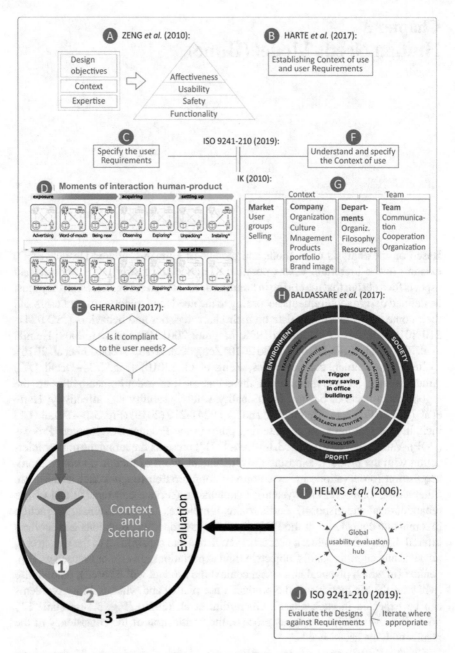

Fig. 3.1 Structure of essential elements of the HUNE model. *Source* (Unruh 2020; Unruh and Canciglieri Junior 2020)

person (used by whom?); (iv) environment and context (used where?). In addition to these authors, Khalid (2006) presents a framework that includes affectivity (intuitive and experimental system with responses of emotions, feelings, and attitudes), in addition to the cognitive aspect (knowledge, meaning and beliefs), as an essential element of the user, and the aspects of context (objectives and activities) and society (norms, habits and trends). Wang and Chen (2011) argued that the foundation of a design is the analysis of human needs and the definition of its requirements, which needs to be done in the initial stages of a PDP.

Thus, the first stage of Human-Centered Design (HCD) should consider the human characteristics (physical, cognitive, and organizational), the objectives, the moments of interaction with product and ergonomics. The context of use and user experience can help in the early stages of a design to achieve a right solution in order to human needs (Chamorro-Koc et al. 2009; Bevan et al. 2015). Therefore, the context has a direct connection with the audience and the user and is relevant to the PDP.

Shluzas and Leifer (2014) explored the importance of context in the workplace, Van Kujik (2010) (Fig. 3.1—Detail "G") argues that the context of aspects must be taken into account before the PDP, especially the market aspects (user groups and sales), company (organization, culture, management, product portfolio and brand image) and departments (organization, philosophy and resources).

Zeng et al. (2010) (Fig. 3.1—Detail "A") also presents a list of contextual factors and human factors to be considered in the PDP, which includes: (i) contextual aspects (design objectives, context and expertise); (ii) ergonomic design values (functionality, safety, usability and affectivity); (iii) seven factors of creativity and ergonomics (novelty, flexibility, aesthetic appeal, interactivity, commonality, simplicity and personalization); and (iv) aspects of value assessment (perceived attractiveness, perceived utility, perceived ease of use, satisfaction, loyalty and profitability).

Harte et al. (2017) (Fig. 3.1—Detail "B"), ISO 9241-210 (2019) (Fig. 3.1—Detail "F") and Baldassare et al. (2017) (Fig. 3.1—Detail "H") also include in their proposals the consideration and understanding of aspects of the context, including social, environmental, and economic aspects for the HCD.

These combined elements can conceptually form a cell (according to the dictionary, a cell is one of several compartments that are part of a whole, or biologically it is the basic structural unit of all organisms), because for a human-oriented PDP a basic structure should also be considered, so that the cell will be the base reference of the model structure, as illustrated in Fig. 3.2, with these elements represented in a dynamic and non-linear way, as well as what are the human factors.

The model name *"HUNE"* originated from English language: HU—Human and NE—Needs. Moreover, consists of eight elements, six of which are human aspects, which can be analyzed or evaluated, a list of requirements and the types of HCD and UCD methods that can assist in these analyses and assessments:

(1) *Human needs*—human needs are the basis of the model because new products are created to meet human needs (Bennett and McWhorter 2019), are the focus of HCD design (Gherardini et al. 2016) and are related to sustainable development (Balyeijusa 2019).

(2) *Human being*—it is essential to consider the human being, discovering who he is first because his characteristics affect his behaviors and needs; this element

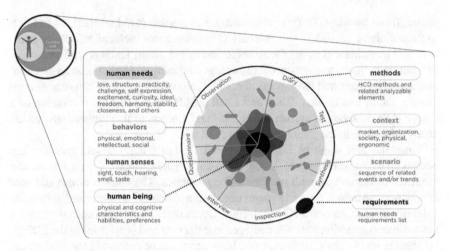

Fig. 3.2   Proposed *HUNE* model. *Source* Unruh (2020)

considers bodily aspects such as anthropometry and biomechanics (Iida 2005), cognitive system, individual differences and preferences (Ponciano et al. 2014) since the theory of product attachment suggests that people develop an attachment to the product for helping them in the construction or maintenance of the aspect of self-identity (Orth et al. 2018);

(3)   *Human senses*—the senses are part of the human being (Ponciano et al. 2014) and it was decided to distinguish them in a separate element, because they are considered the human interface for interaction with the world and outside things and can interfere with some design definitions.

(4)   *Behaviors*—all moments of human-product interaction (Van Kujik 2010), ergonomic aspects related to actions, such as usability and functionality (Zeng et al. 2010), social behavior and emotions (Ponciano et al. 2014).

(5)   *Context*—it is an important starting point of a design and crucial to consider the user because his behaviors are directly linked to the system and the context (Szabluk et al. 2019), includes all aspects related to the context of use (Zeng et al. 2010; Harte et al. 2017; ISO 9241-210 2019), market and company context (Van Kujik 2010), environment, society and profitability (Baldassare et al. 2017);

(6)   *Scenario*—the scenario is a description of what users do while using something (Lee et al. 2011) and can be related to human behaviors, product behaviors, trends, and other non-human factors, that is, it links several the previous elements, in addition to social, political, environmental and market aspects.

(7)   *Requirements*—based on Gherardini et al. (2016), it is recommended to build a list of requirements as a synthesis of all human information gathered at the beginning of a design, with a level of importance of each requirement to be evaluated during the development of the design. Thus, the human information that obtained requirements can be verified in the evaluation phase, helping to define whether the design meets human needs and can move forward in the

Table 3.1 HCD and UCD methods and possible human aspects that can be analyzed

HCD and UCD method types							
Element	Questionnaire	Interview	Observation	Diary	Test	Inspection	Synthesis
Human being	✓	✓	✓	✓			
Human needs	✓	✓	✓	✓			
Sensors	✓	✓	✓	✓	✓	✓	
Behaviors	✓	✓	✓	✓	✓		
Context	✓	✓	✓	✓	✓	✓	
Scenario	✓	✓	✓	✓	✓	✓	✓

Source Unruh (2020)

process or whether the design should be revised in any respect. It is illustrated as a receiver in the cell because it is what generates the communication between human aspects and the product design.

(8) *Methods*—the methods that can be applied for the analysis and evaluation of human aspects during the PDP can be classified into 10 main categories of HCD and UCD methods (Unruh and Canciglieri Junior 2020), of which 7 can be considered for analysis and evaluation in the HUNE model; since the generation of ideas and prototyping are more focused on strict development, so that analysis and evaluation can be applied, and the "other" methods can be applied together with those of analysis and evaluation as complements.

These categories were analyzed and classified, as they can be used for each element of information in the HUNE model, as shown in Table 3.1. In Fig. 3.2, these methods appear around the cell because they are the way to access human information for design development. Products and the line from each method represent the depth range in which they can be analyzed.

The structure of the HUNE model serves as a support for the moments of analysis (initial stages of a design) and evaluation (more advanced stages of a design) of the iterative design cycle, as shown in Fig. 3.3. The cell was designed to support the PDP analysis and evaluation steps, as a guide to help understand which human aspects should be related to the design, assist in choosing and conducting research methods, analysis, and evaluation of HCD, and verifying the adequacy of the design to human needs.

Figure 3.4 presents each element of the HUNE model, and the application process is represented by details "A", "B", "C" and "D", and will be detailed in the next items.

3.1 Definition of Human Aspects

The first step in defining human aspects (Fig. 3.4—detail "A") consists of the hypothetical selection of which aspects of the elements *"human being"*, *"senses"*,

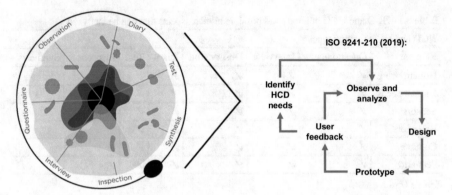

Fig. 3.3 Connection of the HUNE model with the iterative process of ISO 9241-210 (2019). *Source* Unruh (2020)

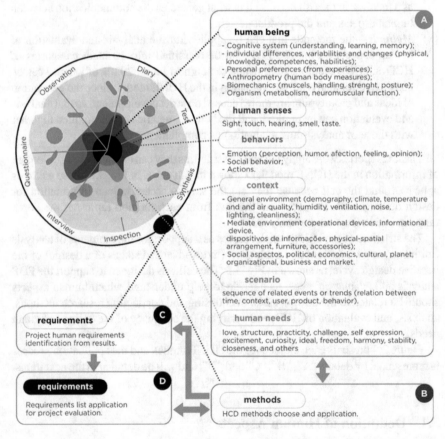

Fig. 3.4 Detailing the elements for application of the HUNE model. *Source* Unruh (2020)

"*behaviors*", "*context*", "*scenario*" and "*human needs*" should be considered conceptually in the product design process. Since the element "human being" contains some characteristics of the profile of the people that one wishes to serve as a focus on the PDP (users, consumers, or target audience), related to their mental and physical shape. Being mental, according to Ponciano et al. (2014), the cognitive system (mental processes, including understanding, learning and memory), individual differences, variability and changes (people's differences in physical aspects, knowledge, skills and abilities), preferences (personal preferences generated by experiences), motivation (impulses to achieve something), emotion (perception, mood, affection, feeling and opinion), and social behavior.

Motivation is directly related to human needs and, therefore, this item is covered in the element "*human needs*". Emotion and social behavior are the results of cognitive processes, composing the item "behaviors". Finally, for physics, according to Iida (2005), aspects of anthropometry (measurements of the human body), biomechanics (muscles, movements, handling, strength, and posture), and organism (neuromuscular function and metabolism) must be considered, in addition to sensors, but these will be considered separately.

The element "*senses*" refers to all human sensors (vision, hearing, smell, taste and touch) since the senses are how a product brings information to people, which are used cognitively for their decision-making decisions and actions, including understanding, concepts, emotions, thoughts and all types of behavior (Iida 2005; Hekkert and Schifferstein 2008; Schifferstein and Spence 2008). It is essential to understand which meanings are related to human-product interaction to develop a pleasant product that is easy to understand, use and interact with the product.

The element "*behaviors*" is related to human actions, these actions are interactions with something or someone, through the senses, which may include emotions, physical behaviors, decisions, and thoughts (Lai 2014). For the specification of this element, the items: actions, social behavior, and emotions (perception, humor, affection, feeling and opinion) will be used, according to Ponciano et al. (2014). The positive feelings that can be expected to be generated through human interactions with a product, presented in the *Geneva Emotions Wheel* (Scherer 2005) are: pride, exaltation, happiness, satisfaction, relief, hope, interest, and surprise.

The element "*senses*" refers to all human sensors (vision, hearing, smell, taste and touch) since the senses are how a product brings information to people, which are used cognitively for their decision-making decisions and actions, including understanding, concepts, emotions, thoughts and all types of behavior (Iida 2005; Hekkert and Schifferstein 2008; Schifferstein and Spence 2008). It is essential to understand which meanings are related to human-product interaction to develop a pleasant product that is the "context" element is essential to understand all the factors of the system in which a product is inserted, which undoubtedly influence people's interaction, understanding and needs, which can affect behaviors and design definitions. Gomes Filho (2010) presented a diagram created by ergonomists with a focus on human–machine interaction, where the context includes the general environment (lighting, ventilation, temperature and air quality, noise, humidity, cleanliness, rain, wind, others) and the mediated environment (operational devices, information devices, physical-spatial arrangement, furniture, equipment, accessories, others). In contrast, Van Kujik (2010) includes market, business, and organization aspects in the context, while Baldassare et al. (2017) include environmental, economic, and social aspects. As the context can include a multitude of aspects that influence human-product interaction, all of these, presented by the authors, will be considered and more "others", which may vary according to the design.

The "*scenario*" element consists of the systemic relationship between the elements previously described, which together form specific situations of use or occasions that influence the use of the product (including time, context, user, product, and behavior) (Dong and Liu 2018), in addition to context-related trends. This element requires a complex and systemic analysis between all elements to identify situations of interaction.

The element "*human needs*" represents the cytoplasm of the cell. In the HUNE model, this element permeates all the others and is what we want to understand in greater depth to address a design-oriented to the human being. There are several theories and definitions of human needs, from the classic presented by Maslow (1971), in addition to more recent studies, such as, for example, Chulef et al. (2001). For the sake of objectivity, in this book, the same categories as Alharti et al. (2018) proposed in their needs identification model, with 12 categories, based on the hierarchy of Ford and Maslow (love, ideal, structure, stability, practicality, challenge, self-expression, excitement, curiosity, freedom, harmony, and closure), plus an "other" item, where the needs of other references can be considered, such as that of Chulef et al. (2001), or more specific needs related to the design, since these mappings are quite a generalist.

Therefore, all these items and aspects of the human elements should be considered as a guide to assist in their identification and analysis in a PDP. However, they are only general indications and, therefore, they can vary and or have more specific developments, according to the specific design in which the model is applied, that is, some aspects of the human elements can be revised according to the needs of the design. Furthermore, to complement these human aspects, several other items, principles, parameters, or guidelines existing in the areas of HCD, UCD, Ergonomics, Usability, and UX can be included.

3.2 Definition and Application of Methods of Analysis and Evaluation of Human Aspects—HCD

The definition and application of the methods of analysis and evaluation of human aspects (HCD) is the second step (Fig. 3.4—detail "B"), where the aspects selected in the first step are thoroughly analyzed, and the hypotheses of the relevant human aspects verified in the design are true and delve into the respective human information and its interaction with the product, context and scenario.

Classic and recent references (Nielsen 1993; Jordan 1998; Hom 1998; Maguire 2001; Bevan 2003, 2009; Heinilä et al. 2005; Cybis et al. 2007; Roto et al.

2010; Cayola and Macias 2018), as well as some websites (usabilidade.gov; designkit.com; ucdtoolbox.com; usabilityplanner.org), present several HCD methods for this purpose, that is, to analyze and evaluate human aspects in their interaction with products.

The research developed by Unruh and Canciglieri Junior (2020) found 89 different methods that were categorized into 10 types of methods, of which 7 can be considered for analysis and evaluation in this model:

1. *Interview*—question-based methods, applied in person (in person or remotely), require direct interaction of participants with experts. Since it is based on conversation, it is excellent for the application of open questions, because it allows exploration of the answers, generating more deep data for analysis and is more suitable for qualitative analysis.

2. *Questionnaire*—methods based on questions employing forms, which can be filled out in person or remotely, not requiring the accompaniment of specialists during their distribution, only in planning and analysis, being able to reach a large number of answers, allowing quantitative analyses. The questions are usually about opinions, experiences, perceptions and or emotions.

3. *Observation*—methods that allow analyzing what people do (their behavior). Seeing people's ways of acting allows analyzing reality, effectiveness, and efficiency, as well as identifying needs. Its origin is in the methods of anthropology and requires an extensive view to seeking to understand the observed reality, motivations, and objectives. A critical view can analyze the situation in detail, identifying aspects that interfere with the situation that can cause problems, accidents, or disappointments and from that, it is possible to identify design opportunities.

4. *Diary*—methods in which users use the product for a period in its actual context. During this time, participants fill out a diary with information about the use and results of using the product. This method is efficient for analyzing contextual aspects, which can only be identified in real use and aspects that only appear with prolonged use. It is indicated to be applied at the beginning of the design with equivalent products, or in advanced stages, with the final product or functional high-fidelity prototypes.

5. *Test*—structured observation methods conducted by an expert moderator. It is complemented by other methods and analyses the usability metrics (effectiveness, efficiency, and satisfaction), according to the ISO 9241-210 (2019) standard. It is planned through a script with tasks for the participant to conduct according to the objectives of using the product, features, and research objectives, simulating real situations of use that make sense to the participants. There are particular methods and methods with more than one user using the product under analysis, and they can be carried out in a controlled environment or a real context. These methods identify the difficulties of use and the understanding that people may have with the product.

6. *Inspection*—methods applied by specialists without user participation, based on scientific, legal, or organizational determination documents, that is, checklists

to inspect the suitability of the product for pre-defined requirements. It can be done with the product at any stage of the design, from initial sketches of the product concept, low-fidelity prototypes or even with the final product or similar products.

7. *Synthesis*—methods that synthesize information collected in several other research, analysis, and evaluation methods, in addition to using the repertoire of those involved, bibliographic research or available product data. Most of them are visual and allow a systemic view and relations of the analyzed product, which helps during the process because the understanding of the information for the development is facilitated.

3.3   Definition of the List of Human Needs Requirements

The definition of the list of human needs requirements is the third step (Fig. 3.4—detail "C"), where the data resulting from the application of HCD analysis methods (with in-depth information about human characteristics, meanings, behaviors, context, scenario, and needs) are transformed into human design requirements.

It is important to have a keen eye, perception and experience in HCD to make this list because it must represent all information from the initial research and analysis of the design focused on the human, clearly and comprehensively, highlighting the identified needs, and without limiting the design possibilities, that is, the requirements cannot indicate specific solutions, but guidelines related to the identified human needs. They must, therefore, be attributes or qualities that the product needs to have or specific problems that it must solve.

These requirements are essential in a product design, help to guide design definitions and decisions, consistently, since they must be based on research and analysis, and applied throughout the PDP. The formatting of HCD analyses in a list of requirements facilitates the consideration of human needs, in an objective and brief material, avoiding that the design team must revisit dense data during the PDP.

3.4   Assessment of Human Needs Requirements

The assessment of human needs requirements is the fourth and last step (Fig. 3.4—detail "D"), and it can be applied more than once in the PDP, and always in conjunction with the HCD methods (Fig. 3.4—detail "B"), being an essential part of the iterative design process. In it, the design is evaluated through the list of human needs requirements to verify if the developed product is meeting them.

Identifying people's needs and requirements is essential for the design of the design (Boztepe 2007). Gherardini et al. (2016) presented a model of criteria for product decision making based on the needs of users and designers. This conceptual approach was incorporated into the HUNE model, assisting in decision-making when it is verified whether the product is suitable for human needs and whether it can be continued in the development process, or whether something must be revised and redone in the design, supporting the iterative design process.

To apply this decision-making process to adapt the design to human needs through the list of requirements, it is necessary to contain at least one requirement related to each element of the HUNE model (human being, senses, behaviors, context, scenario, needs because, as previously identified, all these elements are essential in the analysis of human-product interaction. Only then will it be guaranteed that the minimum of each human aspect related to its interaction with a product is being considered.

The list of requirements must be applied to employ inspection methods (by specialists in HCD or similar area) or by other methods of HCD that allow evaluating the use of the product through human interaction (tests, observation, diary, questionnaires or interviews), (Fig. 3.4—detail "B"), in initial stages (conceptual design—ideas and concepts), intermediaries (strict development—low fidelity prototypes) and final (detailed design—high fidelity prototypes) of PDP.

For the HUNE model, this assessment must assign a score between 0.0 and 1.0 to each requirement. That is if the value "0.0" referring to non-compliance with the requirement and "1.0" to the total compliance with the requirement. In this way, it is possible to determine the value of the "Human Needs Adequacy Indicator (*ANHD*)" which represents whether the design developed is oriented towards human needs or not. Equation (3.1) and Table 3.2 illustrate the calculation methods and the evaluation parameters in 4 percentage ranges (excellent, reasonable, fair, and low), respectively. So, if the *ANHD* value is in the 90–100% range, it will mean that the product is human-oriented. If it is between 70 and 89%, it means that the product will be partially human-oriented, thus being able to proceed with the development, however, preferably, with improvements in the requirements of lower scores. If the *ANHD* range is between 60 and 69%, the product will not be human-oriented and will only be able to proceed with corrections. Furthermore, if the range is less than 60%, the product will certainly be rejected and cannot continue in the development process.

$$ANHD = \frac{\sum \text{Requirements of human need used}}{\text{Number of defined human needs requirements}} 100[\%] \qquad (3.1)$$

where:

- *ANHD* is the value of the indicator of the adequacy of human needs in the design.
- "$\sum$ *Requirements of human need used*" is the sum of the values referring to the requirements of human needs used in product development.
- "*Number of defined human needs requirements*" represents the number of human needs requirements defined at the beginning of the design (Fig. 3.4—detail "C").

Possibilities of applying the HUNE model can be used in unitary form or in conjunction with more cells for each stakeholder who wishes to meet in a design and

Table 3.2 *ANHD* evaluation parameters

Percentage (%)	Rating	Description
90–100	Very good	The product is approved for adaptation to human needs and can proceed to the next stage of the PDP
70–89	Good	The product is partially adapted to human needs and can proceed to the next stage of the PDP, with adjustments or without adjustments, if the design team defines it as such
60–69	Regular	The product is not suitable for human needs and can only go ahead with changes and a new assessment of requirements
< 60	Bad	The product is not suitable for human needs and cannot proceed to the next stage of the PDP. The previous or previous steps must be reviewed and reapplied

can be applied from the initial stages of the design for the analysis of opportunities considering the marketing, product use, after-sales and disposal, or even in more advanced stages, in order to seek improvements in the products and new business opportunities (Fig. 3.5).

Since the human aspects are always the same, but vary according to the profile and types of interaction with a product, the application of more HUNE cells makes the model focused on the HCD, because it can consider all the human profiles that a design desires to answer.

However, this research did not contemplate the application of the model with different stakeholders, only the final "users" of the product, because they end up interacting with the product for a longer time and have a major influence on the return to the business. Besides, because they were experimental cases, there was no

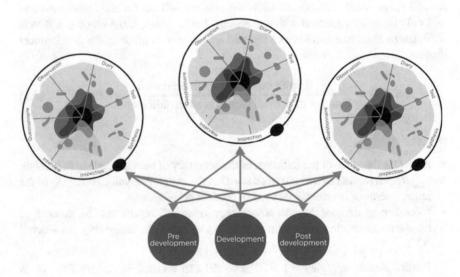

Fig. 3.5 List of several *HUNE* cells for different stakeholders. *Source* Unruh (2020)

access to the company to be able to make the application with people who interact with the product life cycle in the company, as well as there was no monitoring of the entire product life cycle, with the market launch, use and disposal, where it would be possible to include the people who interact at these different times.

In order to assist in the application of the selection of human elements and the selection of HCD methods, a checklist document was created so that the design team can mark the respective aspects of the elements that will be investigated and considered, as well as the methods that will be applied to carry out the analyses and evaluations of these aspects.

3.5 Discussion About Results Obtained from the Proposed Hune Model

Despite the proposed HUNE Model being the product of a deep analysis, and well evaluated by the professionals who applied it, a reanalysis was necessary, seeking a new perspective on some aspects of the model related to existing studies and their relevance.

When redoing the analysis of the literature review, it was possible to verify that there are several models, methods and structures of PDP or Product Design that are very consistent, detailed and grounded, both in scientific articles (especially Van Kujik 2010 and Zeng et al. 2010) as well as in market references (IDEO 2009), in addition there are online platforms that indicate the application of HCD and related methods best suited for each design situation. When analyzing more critically the application of the proposed model, which will be presented in detail in Chap. 4, the specific steps of the business strategy and the restricted development of the product were adequately detailed, and the main contribution of the proposed PDP model was in the analysis and evaluation of human aspects during the PDP, that is, in the central core of the model.

The "analysis" in the model refers to the activities carried out at the time of pre-development in the PDP, where the design opportunities and needs are analyzed and, eventually, during the development, to analyze possibilities of solutions, while the "evaluation" refers to the estimation of the design solutions proposed during development, from the conceptual solution to a high-fidelity prototype, with respect to their suitability for human needs.

It was also possible to notice that despite the existence of several PDP studies oriented to the human being, there was a lack of clarity regarding which human aspects to consider and how to assess their needs in their human-product relationships, and in this sense the core of the preliminary model turned out to be quite useful. Therefore, there is a real need related to the study, but a little different from what was initially proposed. Therefore, the proposed HUNE Model developed is based on all the research and study presented but seeks to emphasize the core of the model initially proposed, and its relational structure with other PDPs existing in the literature.

References

Alharthi R, Guthier B, El Sadik A (2018) Recognizing human needs during critical events using machine learning powered psychology-based framework. IEEE Access 6:58737–58753

Baldassare B, Calabretta G, Bocken N, Jaskiewicz T (2017) Bridging sustainable business model innovation and user-driven innovation: a process for sustainable value proposition design. J Clean Prod 147:175–186

Baleijusa SM (2019) Sustainable development practice: the central role of the human needs language. Soc Change 49(2):293–309

Bennett EE, McWhorter RR (2019) Social movement learning and social innovation: empathy, agency, and the design of solutions to unmet social needs. Adv Dev Hum Resour 21(2):224–249

Bevan N, Carter J, Harker S (2015) ISO 9241-11 revised: what have we learned about usability since 1998? Hum-Comput Interact Part I, HCII 2015:143–151

Bevan N (2003) UsabilityNet methods for user centered design. Human-computer interaction: theory and practice. Proc HCI Int 1(1):434–438

Bevan N (2009) Criteria for selecting methods in user-centered design. Workshop paper I-USED

Boztepe S (2007) Toward a framework of product development for global markets: a user-value-based approach. Des Stud 28:513–533

Cayola L, Macías JA (2018) Systematic guidance on usability methods in user-centered software development. Inf Softw Technol 97:163–175

Chamorro-Koc M, Popovic V, Emmison M (2009) Human experience and product usability: principles to assist the design of user-product interactions. Appl Ergon 40:648–656

Chulef AS, Read SJ, Walsh DA (2001) A hierarchical taxonomy of human goals. Motiv Emot 25(3):191–232

Coelho DA (2010) A method for user centering systematic product development aimed at industrial design students. Des Technol Educ Int J

Cybis W, Betiol AH, Faust R (2007) Ergonomia e Usabilidade: conhecimentos, métodos e aplicações. Novatec Editora, São Paulo

Dong Y, Liu W (2018) Research on UX evaluation method of design concept under multi-modal experience scenario in the earlier design stages. Int J Interact Des Manuf 12:505–515

Gherardini F, Renzi C, Leali F (2016) A systematic user-centered framework for engineering product design in small- and medium-sized enterprises (SMEs). Int J Adv Manuf Technol 91:1723–1746

Gomes Filho J (2010) Ergonomia do objeto: sistema técnico de leitura ergonômica. 2nd ed. Escrituras Editora, São Paulo

Harte R, Glynn L, Rodríguez-Molinero A, Baker PMA, Scharf T, Quinlan LR, Ólaighin G (2017) A human-centered design methodology to enhance the usability, human factors, and user experience of connected health systems: a three-phase methodology. JMIR Hum Fact 4(1)

Heinilä J, Strömberg H, Leikas J, Ikonen V, Iivari N, Jokela T, Aikio KP, Jounila I, Hoonhout J, Leurs N (2005) User centered design: guidelines for methods and tools. The Nomadic Media Consortium

Hekkert P, Schifferstein HNJ (2008) Introducing product experience. Product Experience. Elsevier

Hom J (1998) The usability methods toolbox handbook, 1998. Available in: http://jthom.best.vwh.net/usability/usable.htm. Accessed in: 22nd Sept 2018

Ideo (2009) Human centered design toolkit. Available in: https://www.designkit.org/methods. Accessed in: 3rd Mar 2020

Iida I (2005) Ergonomia: Projeto e Produção. 2nd ed. Edgard Blücher, São Paulo

ISO 9241-210:2019 (2019) Ergonomics of human-system interaction. Part 210: Human-centred design for interactive systems

Jordan PW (1998) An introduction to usability. Taylor & Francis Ltda, London

Karlsson C, Åhlström P (1996) Assessing changes towards lean production. Int J Oper Prod Manage 16(2):24–41. https://doi.org/10.1108/01443579610109820

Khalid HM (2006) Embracing diversity in user needs for affective design. Appl Ergon 37:409–418

Lai Y-C (2014) Emotion eliciting in affective design. In: International conference on engineering and product design education, University of Twente, The Netherlands, 4–5 Sept 2014

Lee KI, Jin BS, Ji YG (2011) The scenario-based usability checklist development for home appliance design: a case study. Hum Factors Ergonomics Manuf Ser Ind 21(1):67–81

Maguire M (2001) Methods to support human-centered design. Int J Hum Comput Stud 55:587–634

Maslow AH (1971) The farther reaches of human nature. Viking, New York

Nielsen J (1993) Usability engineering. Morgan Kaufmann, San Diego

Orth D, Thurgood C, Van Den Hoven E (2018) Designing objects with meaningful associations. Int J Des 12(2):91–104

Ponciano L, Brasileiro F, Andrade N, Sampaio L (2014) Considering human aspects on strategies for designing and managing distributed human computation. J Internet Ser Appl 5(10)

Roto V, Lee M, Pihakala K, Castro B, Vermeerem A, Law E, Väänänen-Vainio-Mattila K, Hoonhout J, Obrist M (2010) All UX evaluation methods. All about UX. Accessed 22 Nov 2018 on http://www.allaboutux.org/

Scherer KR (2005) What are emotions? And how can they be measured? Trends and developments: research on emotions. Soc Sci Inf 44(4):695–729

Schifferstein HNJ, Spence C (2008) Multisensory product experience. Product Experience. Elsevier

Shluzas LMA, Leifer LJ (2014) The insight-value-perception (iVP) model for user-centered design. Technovation 34:649–662

Szabluk D, Berger AVF, Capa A, De Oliveira MF (2019) Design de experiências aplicado à pesquisa: um método exploratório de pesquisa centrada no Usuário. Human Fact Des 8(15):98–113

Unruh GU, Canciglieri Junior O (2020) Identifying and classifying human-centered design methods for product development. Hum Syst Eng Des II, Adv Intell Syst Comput 1026:435–455

Unruh GU (2020) Modelo de análise e avaliação de necessidades humanas para o desenvolvimento de produtos—HUNE. PhD Thesis—Systems and industrial engineering postgraduate program. pontifical Catholic University of Paraná, Curitiba, Brazil

Van Kujik J (2010) Managing product usability: how companies deal with usability in the development of electronic consumer products. Thesis (Doctorate)—Delft University of Technology, Faculty of Industrial Design Engineering. Netherlands

Wang CH, Chen RCC (2011) A MPCDM-enabled product concept design via user involvement approach. Concurrent Eng Res Appl 19(1)

Zeng L, Proctor RW, Salvendy G (2010) Creativity in ergonomic design: a supplemental value-adding source for product and service development. Hum Factors 52(4):503–525

Chapter 4
Application of the Human Needs Model (Hune)

Based on the concepts found in the literature review and in the design and construction of the HUNE Model, this chapter presents six (6) experimental cases. These experiments were carried out using consumer products as they enable more objective solutions than service products. However, the authors believe that the application of the model is also suitable for service products, and this will be the subject of future exploration. The application of the 6 experimental cases had the collaboration of 26 professionals in the area of product development during a period of three months. These professionals were chosen because they put up with, research and work professionally in product development.

4.1 Application 1—Shampoo Packaging for Babies

The first application was in the development of a shampoo package for babies, the main target audience identified was women mothers, according to questionnaires applied by the team with 57 responses. As illustrated in Fig. 4.1, in the first step (Fig. 4.1—detail "A"), there are the human aspects of each element of the model, which were taken into account in the analysis, considering the relationship of people with the baby shampoo.

The second step (Fig. 4.1—detail "B") contains the choice and some results of the application of HCD methods for the initial analysis and definition of requirements. The methods chosen by this team and the main results were:

- *Questionnaire*—it was chosen because it allows the quantitative analysis of profile and preferences aspects. The team obtained 57 responses, concluding that the quality of the product is the biggest factor in the purchase decision, the shampoo

The original version of this chapter was revised: old Figures 4.1 to 4.11 were replaced with the new ones and some unwanted references were deleted. The correction to this chapter is available at https://doi.org/10.1007/978-3-031-12623-9_5

G. Unger Unruh and O. Canciglieri Junior, *Human Needs' Analysis and Evaluation Model for Product Development*, https://doi.org/10.1007/978-3-031-12623-9_4

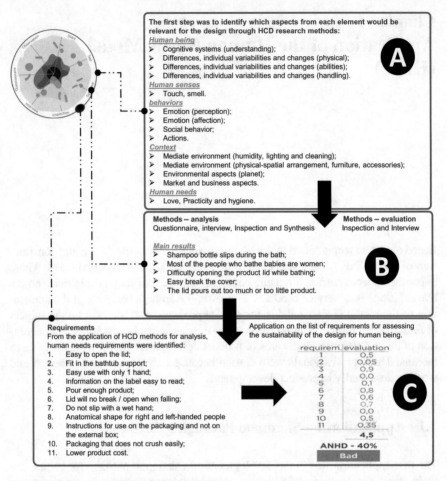

The first step was to identify which aspects from each element would be
relevant for the design through HCD research methods:
Human being
➤ Cognitive systems (understanding);
➤ Differences, individual variabilities and changes (physical);
➤ Differences, individual variabilities and changes (abilities);
➤ Differences, individual variabilities and changes (handling).
Human senses
➤ Touch, smell.
behaviors
➤ Emotion (perception);
➤ Emotion (affection);
➤ Social behavior;
➤ Actions.
Context
➤ Mediate environment (humidity, lighting and cleaning);
➤ Mediate environment (physical-spatial arrangement, furniture, accessories);
➤ Environmental aspects (planet);
➤ Market and business aspects.
Human needs
➤ Love, Practicity and hygiene.

A

Methods – analysis   **Methods – evaluation**
Questionnaire, interview, Inspection and Synthesis   Inspection and Interview

Main results
➤ Shampoo bottle slips during the bath;
➤ Most of the people who bathe babies are women;
➤ Difficulty opening the product lid while bathing;
➤ Easy break the cover;
➤ The lid pours out too much or too little product.

B

Requirements   Application on the list of requirements for assessing
From the application of HCD methods for analysis,   the sustainability of the design for human being.
human needs requirements were identified:
1.   Easy to open the lid;   requirem. evaluation
2.   Fit in the bathtub support;   1   0,5
3.   Easy use with only 1 hand;   2   0,05
4.   Information on the label easy to read;   3   0,9
5.   Pour enough product;   4   0,0
6.   Lid will no break / open when falling;   5   0,1
7.   Do not slip with a wet hand;   6   0,8
8.   Anatomical shape for right and left-handed people   7   0,6
9.   Instructions for use on the packaging and not on   8   0,7
     the external box;   9   0,0
10. Packaging that does not crush easily;   10   0,5
11. Lower product cost.   11   0,35
                                                                          4,5
                                                              ANHD - 40%
                                                                  Bad

C

Fig. 4.1 Application 1—Baby shampoo package

bottle is important, most people have had difficulty opening the product lid or
broke it when opening or falling on the floor, mainly due to the fact of handling
the product while holding the baby.

- *Interview*—it was chosen because it allows a more qualitative and in-depth analysis of experiences. The team applied six interviews of the traditional method,
  where people's experiences and aspects of use were investigated.
- Inspection—it was chosen because it allows for a critical analysis of the market.
  The team applied the analysis of related products, and analyzed five similar products, where it was possible to verify the advantages and disadvantages of the
  format, anatomy of the packaging, and the lid, of the different products.
- *Synthesis*—it was chosen because it allows the creation of a systemic view,
  including actions, context, and usage scenarios. The team applied the user journey
  method in four different scenarios, analyzing the sequence of use and context

aspects, where it was possible to verify several critical points and opportunities for improvement.

From the information collected and analyzed, it was possible to create a list of 11 human requirements for the design, third step (Fig. 4.1—detail "C"), considering at least one aspect of each element of the HUNE model, the requirements were: easy to open lid, fit in baby baths, easy one-handed use, easy-to-read label information, release enough product at the time of use, sturdy lid (won't open or break when dropped), does not slide in the hand while bathing, anatomical shape for right and left-handers, way of use in the packaging, does not wrinkle easily, low cost.

These requirements were then used as part of the design briefing, then an idea for a packaging solution was developed in design and later a low-fidelity prototype was developed, with reuse of materials, where the main focus is on the shape and texture of the bottle, that helps to handle the package, so that it does not slip during the bath, and the lid, which is easier to open than those currently available on the market, and has a spout system for regulated output of product quantity.

On this idea, the fourth step of the HUNE model was applied (detail "C"), the evaluation of the list of requirements in the first product idea (Fig. 4.1—detail "B"). Each requirement was evaluated in the product, by specialists, applying the inspection method, and interviews were also carried out with the target audience of the product, to assess its acceptance and aspects of the requirements.

The analysis of the results of both methods led to the scoring of the requirements, which reached the value of *AHND* of 40%, so it would not be recommended that the product continue the development process, requiring the return of one or more steps, thus generating new other solution alternatives more suited to the requirements imposed in the design.

As this application resulted in a bad factor of *ANHD*, below expectations or prospects, that is, from 11 requirements analyzed in this particular case, only 2 were considered "Good" and "Very good", and these aspects must be reanalyzed and taken into account in the new alternatives of product solution. The aspects that should be reviewed are shown in Fig. 4.2 represented by the symbol "?", asking what could be done in order to improve the "*ANHD* concept" in the return to previous stages. Table 4.1 presents the justification for the evaluation of the requirements and their respective potential for reconsideration, which must be considered to redo the previous steps of the PDP, trying to better the *ANHD* indicator.

Although the *ANHD* indicator was considered bad, most of the results were affected by not having been well defined previously, the constituent materials of the product, high fidelity prototypes, or even more elaborate tests with the definitive materials. Another aspect to be discussed here would be the possibility of a new evaluation, in this way, it would be necessary to go back to the previous steps and review some aspects, or even if the product would follow the determinations of verifying the materials with greater accuracy. In this way, the result can be considered relevant in relation to the application of the HUNE model, because it was possible to verify its potential by guiding and guiding engineers and designers in the decision-making process considering the human requirements in the PDP.

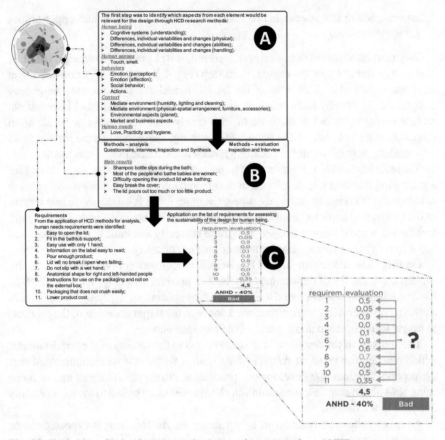

Fig. 4.2 Evaluation of baby shampoo packaging requirements to form ANHD

4.2   Application 2—Bar Soap Packaging

The second application was a bar soap package of superior quality to common soaps found in markets. The main target audience identified was middle class women who want to give someone a gift or take care of their well-being and skin. As illustrated in Fig. 4.3, in the first step (Fig. 4.3—detail "A"), identifying the human aspects of each element of the model that were taken into account in the analysis (considering the relationship of users with the product).

The second step (Fig. 4.3—detail "B") contains the choice and some results of the application of HCD methods for the initial analysis and definition of requirements. The methods chosen by this team and the main results were:

Table 4.1 Baby shampoo packaging requirements with potential for improvement

Requirements (Improvement potential)	Evaluation	Justification of the evaluation	Reconsideration analysis
1	0.5	*Requirement 1*, ease of opening the lid, could not be fully evaluated, because a physical prototype of the lid was not developed, a lid from another product was used in the low-fidelity prototype	Develop the high-fidelity prototype of the lid proposal that opens easily, for testing
2	0.05	*Requirement 2*, fit in the bathtub support, was not met, because the proposed package was too wide	Develop a narrower package, and evaluate the size of the bathtub supports first, to do this properly
3	0.9	*Requirement 3*, easy use with just one hand, was well evaluated because a texture was created on the packaging that promotes hand grip, but it cannot be considered entirely adequate, because this will only be possible to verify in a high-fidelity prototype, with the final material of the product	Develop tests with materials for the product and make a high-fidelity prototype to evaluate the adhesion of the product to the hand during handling with both wet
4	0.0	*Requirement 4*, easy-to-read label information, could not be evaluated because a label solution has not yet been developed at these stages	Develop the packaging label layout, according to requirements 4 and 9
5	0.1	*Requirement 5*, pouring a sufficient amount of product, follows the same reason as requirement 1, could not be evaluated, because a physical prototype of the lid was not developed, and this one has an even lower evaluation, because it depends on the high-fidelity physical prototype with content test	Develop the high-fidelity prototype of the proposed lid content dump engine and evaluate it with the product content

(continued)

Table 4.1 (continued)

Requirements (Improvement potential)	Evaluation	Justification of the evaluation	Reconsideration analysis
6	0.8	**_Requirement 6_**, lid that does not break/open when dropped, also follows the same reason as requirements 1 and 5	Develop the high-fidelity prototype of the cover proposal and test
7	0.6	**_Requirement 7_**, do not slip with a wet hand, will only be possible to evaluate well with a high-fidelity prototype, with the material chosen for the packaging	Develop high-fidelity product prototype and test
8	0.7	**_Requirement 8_**, anatomical shape for right-handed and left-handed users, was met because the packaging does not promote any limitations in terms of use, however it is straight and could be more anatomical	Develop a more anatomical packaging format solution for right and left-handers
9	0.0	**_Requirement 9_**, mode of use in the packaging and not in the external box, was not possible to evaluate, because in these stages a solution for the label and external packaging has not yet been developed	Develop the layout of the label and outer packaging (if any), as per requirements 4 and 9
10	0.5	**_Requirement 10_**, packaging that does not crease easily, will only be possible to fully evaluate with the high-fidelity prototype	Develop high-fidelity product prototype and test
11	0.35	**_Requirement 11_**, lower product cost, could not be evaluated well because there was no focus on product costing	Perform product costing and compare with products on the market

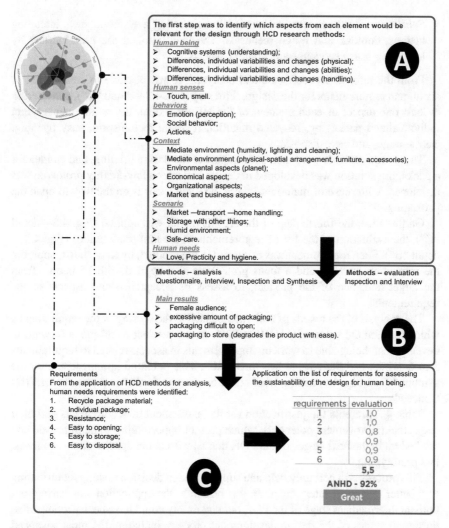

Fig. 4.3 Application 2—Bar soap packaging

- *Questionnaire*—it was chosen because it allows the quantitative analysis of profile and preferences aspects. The team obtained 33 responses, where they analyzed aspects of satisfaction and purchase frequency of the specific type of soap.
- *Interview*—was chosen because it allows a more qualitative and in-depth analysis of experiences. The team applied the focus group method, where information was obtained, mainly related to packaging disposal, price, and storage format.
- *Inspection*—was chosen because it allows for a critical analysis of the market. The team applied the similar analysis method, and analyzed 5 comparable products, where it was possible to verify the advantages and disadvantages of the packaging, including its handling, storage, opening and disposal.

- *Synthesis*—it was chosen because it allows creating a systemic view, including actions, context, and usage scenarios. The team applied the user journey by analyzing usage sequence and context aspects.

From the information collected and analyzed, it was possible to create a list of six human requirements for the design, third step (Fig. 4.3—detail "C"), considering at least one aspect of each element of the HUNE model, the requirements were: individualized packaging, recycled material, resistant to transport, easy opening, easy storage, and easy disposal.

These requirements were then used as part of the design briefing, and an idea for a packaging solution was developed in design and later a low-fidelity prototype was developed, with reuse of materials, where the main focus is on the way to open the packaging.

On this idea, the fourth step of the HUNE model was applied (Fig. 4.3—detail "C"), the evaluation of the list of requirements in the first product idea (Fig. 4.3— detail "B"). Each requirement was evaluated in the product, by a specialist, applying the inspection method, and a focus group was also applied with 15 people from the target audience of the product, to assess its acceptance and aspects of the requirements.

The analysis of the results of both methods led to the scoring of the requirements, which reached the value of *AHND* of 92%, so the product could move forward in development, being able to work on improvements in aspects related to requirements 3 to 6, which had their value less than "1.0". This potential is represented by the symbol "?" in Fig. 4.4, asking what could be done to further improve the "*ANHD* Concept".

Table 4.2 presents the justification for the assessment of requirements and their respective improvement potentials, which present opportunities to be improved and worked on in the next stages of the PDP, that is, with better specifications, analysis, and prototyping.

This process will certainly help and support design decision making, contributing to a better design of later stages. In this research, the application was carried out until an intermediate stage of the PDP, but it certainly could have been accompanied during all stages of the design development process, including the most advanced stages, with new analysis of the requirements through methods with the participation of a number of larger number of users. With this, it is believed that the application of the iterative process offered by the HUNE Model could raise the *ANHD* indicator to the "optimal" concept, that is, between 90 and 100%.

4.3 Application 3—Soap Bar Packaging for Babies

The third application was a bar soap package for babies, the main target audience identified was women as well. As illustrated in Fig. 4.5, in the first step (Fig. 4.5— detail "A"), there are the human aspects of each element of the model, which were

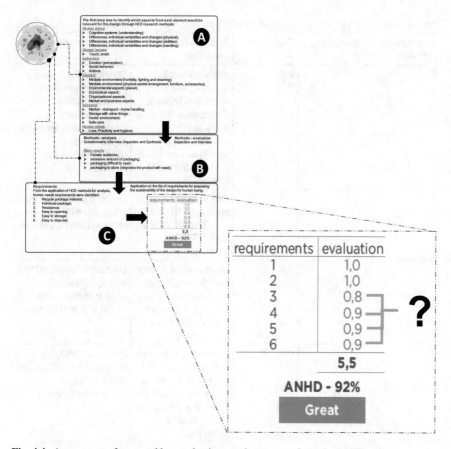

Fig. 4.4 Assessment of soap and bar packaging requirements to form the ANHD

considered in the analysis, considering the relationship of people with the bar soap for babies.

The second step (Fig. 4.5—detail "B") contains the choice and some results of the application of HCD methods for the initial analysis and definition of requirements. The methods chosen by this team and the main results were:

- **Questionnaire**—it was chosen because it allows to analyze quantitatively aspects of profile and preferences. The team obtained 89 responses, where they analyzed profile aspects, use of specific products for babies, sensory aspects, especially smell related to soap, use and disposal of the packaging.
- **Inspection**—was chosen because it allows for a critical analysis of the market. The team applied the similar analysis method, and analyzed 3 comparable products, where it was possible to verify the advantages and disadvantages of the packaging, including its handling and storage.

Table 4.2 Bar soap packaging requirements with potential for improvement

Requirements (Improvement potential)	Evaluation	Justification of the evaluation	Reconsideration analysis
3	**0.8**	**Requirement 3,** resistant, was well evaluated, because the material chosen is theoretically resistant, but a prototype was not made with the real material to be sure	Develop a high-fidelity prototype with the chosen material and assess the material's strength during transport handling and medium-long term storage
4	0.9	**Requirement 4**, easy opening, will also only be right with a high-fidelity prototype	Develop a high-fidelity prototype with the chosen material and test
5	0.9	**Requirement 5**, easy storage, was highly rated because of its size, but would need to be evaluated in real context	Develop a high-fidelity prototype with the chosen material and assess it in real context
6	0.9	**Requirement 6**, easy disposal, was well evaluated, but would also need to undergo a real-life test and an environmental assessment	Evaluate packaging disposal in a real context and with waste management experts

- **Synthesis**—it was chosen because it allows creating a systemic view, including actions, context, and usage scenarios. The team applied usage scenarios and user journey methods by analyzing usage sequence and context aspects.

From the information collected and analyzed, a series of problems with the current use of the product and opportunities for improvement were identified, such as the situation of holding the baby and handling the soap at the same time, the relationship between the bathtub and the soap support, and also the difficulty of transporting soap on trips.

Then it was possible to create a list of 5 human requirements for the design, third step (Fig. 4.5—detail "C"), considering at least one aspect of each element of the HUNE model, the requirements were: the ease of being stacked, enable the olfactory perception, the packaging needs to last as long as the product is used, easy disassembly, made with renewable materials.

These requirements were then used as part of the design brief, and the team developed some alternatives that were evaluated with users, first through images and

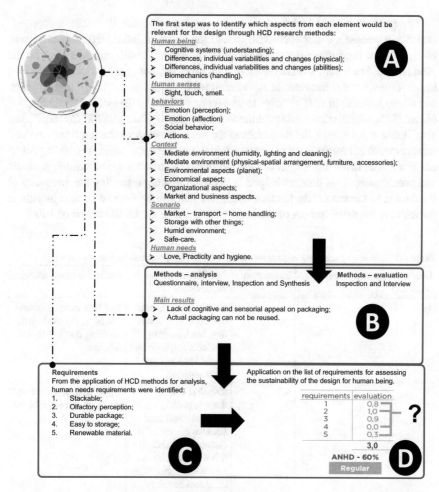

The first step was to identify which aspects from each element would be relevant for the design through HCD research methods:
Human being
➤ Cognitive systems (understanding);
➤ Differences, individual variabilities and changes (physical);
➤ Differences, individual variabilities and changes (abilities);
➤ Biomechanics (handling).
Human senses
➤ Sight, touch, smell.
behaviors
➤ Emotion (perception);
➤ Emotion (affection)
➤ Social behavior;
➤ Actions.
Context
➤ Mediate environment (humidity, lighting and cleaning);
➤ Mediate environment (physical-spatial arrangement, furniture, accessories);
➤ Environmental aspects (planet);
➤ Economical aspect;
➤ Organizational aspects;
➤ Market and business aspects.
Scenario
➤ Market – transport – home handling;
➤ Storage with other things;
➤ Humid environment;
➤ Safe-care.
Human needs
➤ Love, Practicity and hygiene.

A

Methods – analysis
Questionnaire, interview, Inspection and Synthesis

Methods – evaluation
Inspection and Interview

Main results
➤ Lack of cognitive and sensorial appeal on packaging;
➤ Actual packaging can not be reused.

B

Requirements
From the application of HCD methods for analysis, human needs requirements were identified:
1. Stackable;
2. Olfactory perception;
3. Durable package;
4. Easy to storage;
5. Renewable material.

Application on the list of requirements for assessing the sustainability of the design for human being.

requirements	evaluation
1	0,8
2	1,0
3	0,9
4	0,0
5	0,3
	3,0

?

ANHD - 60%
Regular

C **D**

Fig. 4.5 Application 3—Bar soap packaging for babies

an online questionnaire and then with a medium–high fidelity 3D printed prototype and observation. In addition to validating the requirements twice during development, the result was a highly resistant and versatile package, it can be reused to transport the soap on trips or to hang the open package as a support, in the shower itself or on the water handle.

On this idea, the fourth step of the HUNE model was applied (Fig. 4.5—detail "C"), the evaluation of the list of requirements in the first product idea (Fig. 4.5—detail "B"). Each requirement was evaluated in the product, by a specialist, applying the inspection method, and interviews were also conducted with 5 people from the product's target audience, to assess their acceptance and aspects of the requirements.

The product proved to be easy to open and easy to handle while showering, improving people's experience with it, and the value of the AHND indicator was 60%, so some requirements were well met, but others can still improve a lot, so the product could move forward in development, however, with the restriction of serious improvements and refinement in aspects related to requirements 1, 3, 4 and 5, which had values lower than "1.0". This potential is represented by the symbol "?" in Fig. 4.5 (detail "C"), asking what could be done to further improve the "*ANHD* Concept", and also, Table 4.3 presents the justification for the evaluation of the requirements and their respective improvement potentials, which present opportunities to be improved and worked on in the next stages of the PDP, that is, with better specifications, analysis and prototyping. This process helped in decision making regarding the adequacy of the design to human needs, leading to aspects that can be worked on and improved throughout the development of the product to achieve an *ANHD* factor of 100%.

Table 4.3 Bar soap packaging requirements for babies with potential for improvement

Requirements (Improvement potential)	Evaluation	Justification of the evaluation	Reconsideration analysis
1	0.8	*Requirement 1*, ease of stacking, was well met, but how many products can be safely stacked has not yet been assessed	Test how many products can be safely evaluated, avoiding damage to the product
3	0.9	*Requirement 3*, durable packaging, was well evaluated because the material chosen is durable, but it is only possible to verify this factor with complete certainty through long-term use tests	Apply long-term tests to verify product durability
4	0.0	*Requirement 4*, easy storage, could not be evaluated, because no tests were performed in real context	Apply product testing in real context to verify the ease of storage in different related contexts
5	0.3	*Requirement 5*, renewable materials, had a low evaluation because in these stages the specific material has not yet been defined	Analyze possible renewable materials to be used in the product

4.4 Application 4—Hand Moisturizing Cream Packaging

The fourth application was a package of moisturizing hand cream, considering especially women as the target audience. As illustrated in Fig. 4.7, in the first step (Fig. 4.6—detail "A"), there are the human aspects of each element of the model, which were considered in the analysis, considering the relationship between people and the hand cream.

The second step (Fig. 4.6—detail "B") contains the choice and some results of the application of HCD methods for the initial analysis and definition of requirements. The methods chosen by this team were:

- *Questionnaire*—because it allows a broad analysis of the profile and some human aspects and preferences, quantitatively.

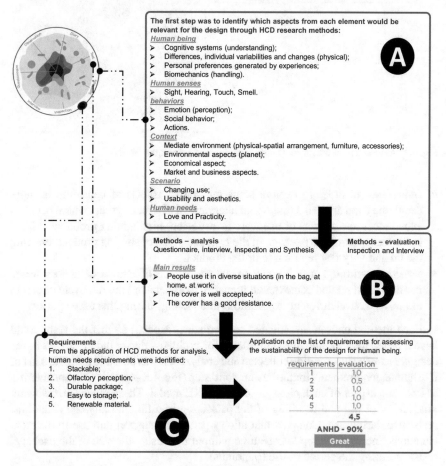

Fig. 4.6 Application 4—Moisturizing hand cream package

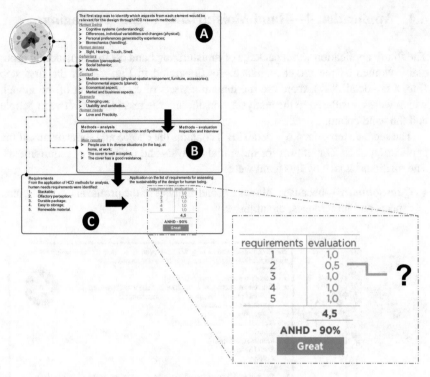

Fig. 4.7 Assessment of cream packaging requirements to form the ANHD

- *Interview*—in order to explore some qualitative aspects of behaviors in more detail, the team applied 3 traditional interviews and a focus group, where people's experiences and aspects of use were investigated, highlighting problems related to the transport of the package in the bag, accidents of spilling product, opening the lid and very fast repurchase of the product.
- *Synthesis*—through user journey, to analyze aspects of actions, behaviors, usage moments and related context, and scenario, to understand the full cycle of cream usage, as well as different expectations and feelings during this usage period.

Each method chosen investigated some distinct aspects, so that the team could have a broad look at human requirements for problems that could be solved in the design. From the information collected and analyzed, it was possible to create a list of five human requirements for the design, third step (Fig. 4.6—detail "C"), considering at least one aspect of each element of the HUNE model. These requirements were: practicality of use and handling of the packaging in different daily use situations, resistance during use. Compact that allows personal transport and use in different situations, increase the repurchase time without increasing the size of the package, resistant during the period of use (6 months).

An idea for a packaging solution was developed in design and later a prototype was developed, where the main focus is on the lid of the packaging, and on the format of fitting the part of the content of the product, and people will be able to buy both separately, reusing the cap for different amounts of product, which makes it possible for each person to buy the most suitable amount for himself, and not have to throw the whole package away every time he buys a new product, thus also generating brand loyalty. In addition, the cover is developed with a more resistant material than the common ones, promoting the resistance requirement.

On this initial idea, the fourth step of the HUNE model was applied (Fig. 4.6—detail "C"), the evaluation of the list of requirements in the first product idea, through a method of inspecting this idea and the prototype (Fig. 4.6—detail "B"). Each requirement was evaluated in the product, by specialists, applying the inspection method, and a questionnaire was also applied with people from the product's target audience, to assess its acceptance and aspects of the requirements.

The analysis of the results of both methods led to the requirements score, which reached the AHND value of 90% (optimal), so the product could move forward in development. However, it would still be possible to improve aspects related to requirement 2, which had its value lower than "1.0". This potential is represented by the symbol "?" in Fig. 4.7, asking what could be done to further improve the *ANHD Concept*".

Requirement 2 refers to the strength of the packaging during the use of the product, and it is an aspect that can only be fully evaluated through a high-fidelity prototype, with the appropriate material of the product, at more advanced stages of the design. Therefore, the design should continue the development process, paying attention to the evaluation of this requirement in more advanced stages, as presented and described in Table 4.4. This process will help and support design decision-making, contributing to a better design of the later steps.

Table 4.4 Cream packaging requirements with potential for improvement

Requirements (Improvement potential)	Evaluation	Justification of the evaluation	Reconsideration analysis
2	0.5	As *requirement 2*, toughness in use, is an aspect that can only be fully assessed through a high-fidelity prototype, with the correct product material, at more advanced stages of the design	Apply the design of the packaging lid to a long-term resistant material (2 to 5 years approximately) and the packaging body to a medium-term resistant material (6 to 12 months) to contact with other products inside a bag, for example, and the use itself. Test the strength of the material with cream inside

4.5 Application 5—Eyelash Mask

The fifth application was an eyelash mask, the main target audience identified was women as well. As illustrated in Fig. 4.8, the first step (Fig. 4.8—detail "A") contains the human aspects of each element of the model, which were taken into account in the analysis, considering the relationship between people and the product. The second step (Fig. 4.8—detail "B") contains the choice and some results of the application of the HCD methods for the initial analysis and definition of requirements. The methods chosen by this team and the main results were:

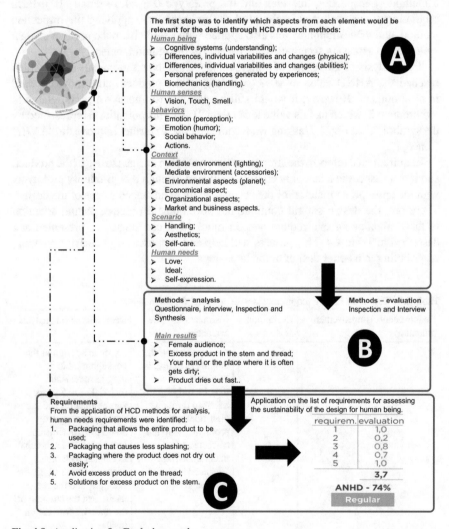

Fig. 4.8 Application 5—Eyelashes mask

- *Questionnaire*—it was chosen because it allows the quantitative analysis of profile and preferences aspects. The team obtained 103 responses, in which they analyzed aspects of profile, use and purchase of masks, experiences and difficulties, where some problems were already identified, such as rapid drying of the product, product remaining on the shaft, which ended up dirtying hands or clothes. of users.
- *Observation*—it was chosen because it allows verifying the situation and the real context, the field observation method was applied with 6 users, observing the environment, and especially the actions of use.
- *Inspection*—was chosen because it allows for a critical analysis of the market. The team applied the similar analysis method and analyzed 5 related products, where it was possible to verify the advantages and disadvantages of the packaging, including its handling and storage.
- *Synthesis*—it was chosen because it allows creating a systemic view, including actions, context, and usage scenarios. The team applied the user journey method by analyzing usage sequence and context aspects in 3 usage scenarios.

From the information collected, it was possible to create a list of five human requirements for the design, third step (Fig. 4.8—detail "C"), considering at least one aspect of each element of the HUNE model, the requirements were: packaging that allow using the entire product, causing splashing, packaging where the product does not dry out easily, avoid excess product on the screw, avoid excess product on the stem.

These requirements were then used as part of the design brief, and the team developed some alternatives and chose one to prototype and evaluate with users. A digital 3D modelling was developed, which was evaluated through a questionnaire, with 72 responses, and then a medium-fidelity prototype of molded and painted polyurethane was made, evaluated through expert inspection and contextual interview, with 10 users using the prototype. and answering evaluation questions. The product presents a solution regarding the cleaning of the rod, in which the rod is retracted and extracted in each use, in a way that dirt stays inside the container, in addition the product content is in a component that is a refill, or that is, it can be purchased separately, and it is made with a flexible material, a type of rubber, which allows you to move the contents, avoiding drying and taking better advantage of the contents of the same until the end.

The value of AHND, resulting from the assessment by questionnaire, inspection and contextual interviews was 74%, where some requirements were well met, but others can still improve, so it would still be possible to improve aspects related to requirements 2, 3 and 4, which had their values less than "1.0". This potential is represented by the symbol "?" in Fig. 4.9, asking what could be done to further improve the "*ANHD* Concept".

Table 4.5 presents the justification for the assessment of requirements and their respective improvement potentials, which present opportunities to be improved and worked on in the next stages of the PDP, that is, with better specifications, analysis, and prototyping. This process will certainly help and support design decision

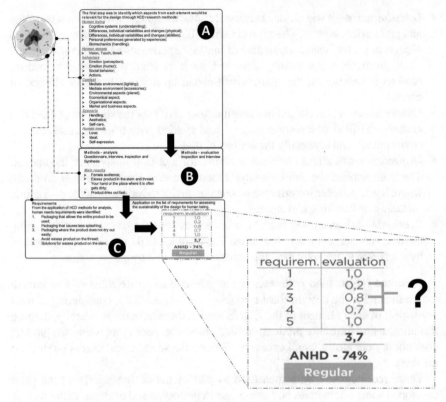

Fig. 4.9 Assessment of mascara requirements to form the ANHD

making, contributing to a better design of the later stages, with new analysis of the requirements through methods with user participation, with the objective of raising the ANHD concept to "optimal", that is, between 90 and 100%.

4.6 Application 6—Refrigerator

The sixth situation was the conservation of food in a domestic environment by means of refrigeration, considering as the target audience of the design middle class couples who live together, without children, with the objective of identifying opportunities for the development of a new domestic refrigerator, for this was not chosen a specific product for initial analysis, but a situation related to the type of product, in order to help in understanding the needs related to it for the development of a new one.

The steps of this application of the HUNE model are illustrated in Fig. 4.10. The first step (Fig. 4.10—detail "A") is the selection of aspects of each element of the model that relate to the situation in focus, which aims to guide the investigation

Table 4.5 Eyelash mask packaging requirements with potential for improvement

Requirements (Improvement potential)	Evaluation	Justification of the evaluation	Reconsideration analysis
2	0.2	*Requirement 2*, packaging that causes less splashes, was not well evaluated because the thread proposed in the solution still has the same shape as current threads, and the content of the product is also the same, so despite the thread getting dirty less because it has a protection and being retractable, it can still cause splashing	Check if there is a possibility to change the shape of the thread to generate less spatter or if it is possible to make any changes to the product content to avoid this
3	0.8	*Requirement 3*, packaging where the product does not dry out easily, was well evaluated because the product has a flexible material that allows the contents to be moved, and a format that avoids the permanence of a lot of air, avoiding drying, but it will only be possible to be sure addition with long-term use tests	Apply long-term use tests with high-fidelity prototypes to verify product dryness
4	0.7	*Requirement 4*, avoid excess product in the thread, was well evaluated because of the shape of the mask and the retractability, but dirt can remain between the product body and the cover/lid	Check the possibility of improving the screw system or the chemistry of the product, to avoid these excesses

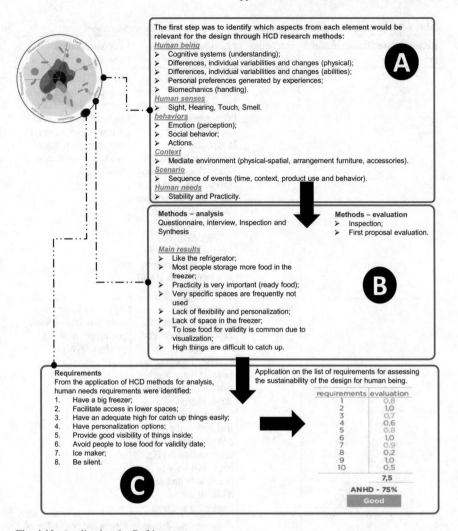

The first step was to identify which aspects from each element would be relevant for the design through HCD research methods:

Human being
➢ Cognitive systems (understanding);
➢ Differences, individual variabilities and changes (physical);
➢ Differences, individual variabilities and changes (abilities);
➢ Personal preferences generated by experiences;
➢ Biomechanics (handling).

Human senses
➢ Sight, Hearing, Touch, Smell.

behaviors
➢ Emotion (perception);
➢ Social behavior;
➢ Actions.

Context
➢ Mediate environment (physical-spatial, arrangement furniture, accessories).

Scenario
➢ Sequence of events (time, context, product use and behavior).

Human needs
➢ Stability and Practicity.

A

Methods – analysis
Questionnaire, interview, Inspection and Synthesis

Main results
➢ Like the refrigerator;
➢ Most people storage more food in the freezer;
➢ Practicity is very important (ready food);
➢ Very specific spaces are frequently not used
➢ Lack of flexibility and personalization;
➢ Lack of space in the freezer;
➢ To lose food for validity is common due to visualization;
➢ High things are difficult to catch up.

Methods – evaluation
➢ Inspection;
➢ First proposal evaluation.

B

Requirements
From the application of HCD methods for analysis, human needs requirements were identified:
1. Have a big freezer;
2. Facilitate access in lower spaces;
3. Have an adequate high for catch up things easily;
4. Have personalization options;
5. Provide good visibility of things inside;
6. Avoid people to lose food for validity date;
7. Ice maker;
8. Be silent.

C

Application on the list of requirements for assessing the sustainability of the design for human being.

requirements	evaluation
1	0,8
2	1,0
3	0,7
4	0,6
5	0,8
6	1,0
7	0,9
8	0,2
9	1,0
10	0,5
	7,5

ANHD - 75%

Good

Fig. 4.10 Application 6—Refrigerator

of these elements through HCD methods, aspects of each element were selected, identified as important in the human-refrigerator relationship, including physical, cognitive, context aspects, etc.

The second step (Fig. 4.10—detail "B") was the choice and application of the "questionnaire" and "diary" HCD methods, as they can be applied to analyze all elements, and the results relied on data from the speeches of the participants. People and observed behaviors, important to compare and analyze in greater depth, context situations and behaviors.

Through the application of the methods, it was possible to identify a series of information about the target audience of the design and its interaction with the situation of preserving refrigerated food in their homes.

The third step (Fig. 4.10—detail "C") was the transformation of the information collected in the second step into a list of 10 human needs requirements of the design, considering at least one aspect of each element of the HUNE model. These requirements were then used as part of the design brief to guide its development.

So, an initial idea for a new refrigerator solution was developed, lower part of Fig. 4.10—detail "B", with 3 lower drawers, which allows the selection of the temperature (to refrigerate or freeze), according to the preferences of each user, with the top for cooling only, with customizable compartments, and with options to purchase accessories for specific uses separately.

On this initial idea, the fourth step of the HUNE model was applied (Fig. 4.10—detail "C"), the evaluation of the list of requirements in the first product idea, through an inspection method (Fig. 4.10—detail "B"). Each requirement was evaluated in the product, only by an expert, reaching an AHND result of 75%, so you can move on in the development, to be then more detailed. With this, the design could continue its development and detailed process. However, observing the values obtained from the requirements presented in Fig. 4.10 and focusing on its development to be successful, it would be interesting to reassess or propose changes to requirements 1, 3, 4, 5, 7, 8 and 10 that have potential for improvement in their design. This potential is represented by the symbol "?" in Fig. 4.11, meaning, or even questioning what could be done in terms of product design to improve the "ANHD Concept".

Table 4.6 presents the justification for the assessment of requirements and their respective improvement potentials. Therefore, all the requirements presented in Table 4.6 have values that present opportunities to be improved and worked on in the next stages of the PDP, that is, with better specifications, analysis, and prototyping.

This entire process supported Designers and Engineers in making design decisions, contributing to a better delineation of later stages. In this research, particularly in the experimental case of the refrigerator, the application was carried out until the initial stage of the PDP, but it certainly could have been followed during all stages of the design development process.

4.7 Final Discussion

The systematic review of the literature developed by Unruh and Canciglieri Junior (2020) was fundamental in the conduct of all the research and analysis, and clearly, the fact that it was systematic, with exploratory scientific objectives, definition of terms, criteria and registration of the process, brought a robust foundation to the study composed of references relevant to the topic addressed, enabling the creation of the HUNE Method.

The research process developed, not explored in detail in this book, was long and ended up going through several stages, initially influenced by an initial problem

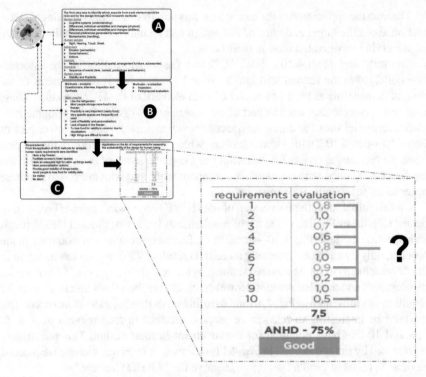

Fig. 4.11 Assessment of refrigerator requirements to form the ANHD

and previous objective, to create a new PDP model, which also influenced the data analysis and references, but after its application and publication attempts, it led to the need for a new analysis of the data, from a different perspective, more open to understanding the real limitations and opportunities existing in the area. This process showed that care must be taken with the perspective from which research is applied, so that it is not influenced, which may lead to an incoherent result.

And this process was also positive, because it made it possible to reanalyze the data, changing the perspective and identifying with greater clarity the real opportunity, to which this thesis could contribute, instead of a PDP model, a model that helps existing PDPs in the analysis and assessment of human needs. And this was only possible because the systematic review of the literature was robust and reapplied for constant updates throughout the development of the research, that is, because a preliminary model was developed and applied, enabling its analysis that led to the proposal of a new refined model. more objective, which was also applied and analyzed in a critical way, and which, after verifying that some aspects were still not consistent with respect to existing studies, led to a reanalysis of all references, applications, processes, and evaluations, in a judicious and changing way. The perspective, which resulted in a new model proposal, finally coherent, clear, robust and with a more appropriate and modern look.

Another key factor in the proposal was the process of creating and evaluating human needs requirements, which contributes to decision making regarding the suitability of products to human beings in a clearer and more systematic way than methods, models and structures existing so far.

Table 4.6 Refrigerator requirements (justification and potential for improvement)

Requirements (Improvement potential)	Evaluation	Justification of the evaluation	Reconsideration analysis
1	0.8	As requirement 1 would be to have a more spacious freezer, and the design proposal includes drawers that can be used as a refrigerator or freezer, each user will have as much space as they want, between one and three drawers, however, if you have to store something very tall that needs to stand, the drawers may not hold	Reassess in more detail whether the designed height can accommodate all types of products that users would store in the freezer, otherwise design one of the drawers with a higher height
3	0.7	As requirement 3 is for the refrigerator to have adequate height to reach the food at the top, the proposed new refrigerator is 1.70 m high (lower than those evaluated in the usage log), but still some people would have difficulty reaching things in the top of the refrigerator	Make the height of the refrigerator even lower (but it can harm taller people), or design some mechanism for the upper shelves, which allows accessing them more easily, bringing them to the front or down
4	0.6	Requirement 4 is to allow the customization of the height of the shelves, the score was "0.6" because in the initial proposal there are some height options, even so they are still limited	Develop a shelving system that allows full customization, in height, depth and, if possible, width

(continued)

Table 4.6 (continued)

Requirements (Improvement potential)	Evaluation	Justification of the evaluation	Reconsideration analysis
5	0.8	Requirement 5 is to allow customization of the location of compartments, and it had a score of "0.8" because the proposed compartments are less limited to the type of content than some current refrigerators, and on the door, there are several possibilities of fitting, but the system still needs to be better thought out to work well	Detail the size or design a system for the door compartments, which allows the adaptation of space as needed (reduce or increase the size of the compartment)
7	0.9	Requirement 7 is to provide good visibility and has almost reached 100% because the drawers help a lot in this regard, but items on the highest shelf can still suffer a little with visibility	Design some mechanism for the upper shelves, which allows accessing them more easily, bringing them forward or down
8	0.2	Requirement 8 had the lowest score, "0.2", which is to prevent people from missing the expiration date, it is a difficult requirement to meet, it would probably have to work with some technology system to help with this, and it could be better developed in the next steps, but it would make the product more expensive	Develop an intelligent system to identify expiration dates of products placed in the refrigerator or enable the user to categorize "zones" inside the refrigerator, by color or by some accessory, which identifies products with shorter and longer expiration dates than the product itself. user can select and organize
10	0.5	Requirement 10, being silent, was evaluated with a partial score because it will only be possible to evaluate effectively through the functional prototype	Develop a quieter engine or soundproofing of the engine, to reduce its noise

Therefore, the proposal of the HUNE model, which helps the insertion of human aspects in existing PDP models, methods, and structures, as well as its organic and flexible form, is current and in fact contributes to the existing limitations and difficulties of HCD application. in the PDP, because it is not static, like previous models, on the contrary, it is dynamic and allows diverse applications and constant updates that are coherent with the current social and technological dynamics.

Even so, the HUNE model still has limitations and opportunities for improvement, the main limitation was to propose a model focused on the "human being", but it was possible to apply the model only in cases focused on the end "user", that is, the person who will use the product. Another limitation was to consider only consumer products in the applications and not service products, which it is believed that the model also meets, however specific applications would have to be made in this context.

Although the model and the study are theoretically located in areas of knowledge that can contribute to social innovation and sustainability, the model did not delve into these aspects nor into related applications and evaluations.

The research identified the lack of detail on how to apply the HCD methods in existing models, and the proposed model, despite helping to conduct which aspects to analyze and evaluate in the application of the methods, also did not go into the detailed specifications of how to apply the methods. Methods of analysis and evaluation itself.

One opportunity would be to improve the decision-making process or conduct which method of analysis or evaluation of HCD to apply, since there are many, and although the model categorizes the methods and indicates what information they can consider, and there are online platforms in the assistance, is still quite broad, there are many possibilities, and this requires specific experiences of experienced professionals who will apply the methods avoiding doubts.

Another factor that can still be improved is the definition of requirements and their evaluation, since the requirements vary in each specific design, and the level of importance of each requirement can also vary, since human beings are complex and some aspects may be more and less relevant, there could be a way to consider these different levels in defining and evaluating human needs requirements.

Further research on the aspects of human elements, which are part of the HUNE model, can still be worked on in future research in the form of a guide, a checklist to consider in the analyses and evaluations, but there is a lot of information inherent to each aspect that can be deepened, with knowledge of areas such as sociology, anthropology and psychology, generating guidelines, recommendations or just important data, in general and even focused on specific types of products.

Finally, the application of meta design tools or something similar can be analyzed and studied that allows questioning, adapting, and reinventing the model itself, since both the human being, society and the PDP are complex systems and can change according to the situation. Specific.

If this research were to be redone, it should start from an initial analysis of the literature review, less influenced by the previous objective of developing a PDP model, expanding the look at the different opportunities, which would lead to the deepening mentioned in the previous paragraphs at the beginning of the research, as well as it could consider references from areas such as sociology and anthropology, which can be contemplated in the present thesis. Another point to be considered is that more visual and systematic documentation of analyzed information could be made, which could clarify the understanding of the analyses and change the perspectives and perceptions of needs, as well as visual documentation and with more information of each element of the HUNE model.

It is clear that knowledge is inexhaustible, that points of view influence the analyses and that there will always be the possibility of new research, new solutions and new proposals, therefore, based on the entire research and analysis process, on the evolution of research and in the evaluations of professionals, students and specialists, of the HUNE model, this thesis brings relevant contributions to the insertion of the HCD in the PDP, however it has limitations and opportunities for future research.

The proposal of the HUNE model was detailed in book, indicating means of application and due details to fulfil its objective of assisting in the moments of analysis and evaluation of human aspects in the PDP. The applications demonstrated clarity and effective conduction of the model application process, through different results, which indicate how the model is flexible and in fact helps in decision making in the design.

In most cases, similar types of analysis and evaluation methods were used, even though they varied in the specific method, there was a large application of questionnaires, interviews, observations, inspection and synthesis, and the types of interview varied in some cases (focus groups, individual, contextual interviews), as well as the summaries (use contexts, usage scenarios and user journey) and inspections (analysis of similar, analysis of requirements), this was because participants were asked to apply a variety of methods that include listening and observing users, because interviews, questionnaires, observations and summaries are more suitable at the time of analysis (pre-development, identification of needs) while inspection is more suitable at the time of the first evaluation (with an initial proposal or low-fidelity prototype), and because they are the most practical methods known to professionals. The journal method was applied in only one case, because it demands more follow-up time, and it was not possible to apply tests, because they demand high-fidelity prototypes.

The applications only contemplated the focus on people who use the product after its consumption, that is, on "users", but it is believed that the model can be applied to different profiles of people who could interact with the product and, this, can be a topic for future studies.

Since the applications only went up to the initial stages of a PDP, it was possible to perceive that in all cases, decisions to improve the identified requirements can still be made for the next stages of the design, in some cases more and in others any less. However, despite not contemplating the PDP until the final stages and launch, the applications were enough to verify the complete process of application of the proposed model. In the continuation of this application, what would happen would be just the repetition of the application of the steps, in an iterative way, until reaching the best result and the final product.

Reference

Unruh GU, Canciglieri Junior O, (2020) A proposed model for analysing and assessing human needs in the product development process. Hum Factors Des HFD 9(18):52–77

Correction to: Human Needs' Analysis and Evaluation Model for Product Development

Correction to:
G. Unger Unruh and O. Canciglieri Junior, *Human Needs'*
Analysis and Evaluation Model for Product Development,
https://doi.org/10.1007/978-3-031-12623-9

In the original version of Chapter 2 and 3, some of the references were omitted. This has now been rectified and the references have been added.

In the original version of Chapter 4, the old Figures 4.1 to 4.11 were not replaced with the new figures and some unwanted references were not deleted. The correction has been updated in the current version.

The updated original versions of these chapters can be found at
https://doi.org/10.1007/978-3-031-12623-9_2
https://doi.org/10.1007/978-3-031-12623-9_3
https://doi.org/10.1007/978-3-031-12623-9_4

Printed in the United States
by Baker & Taylor Publisher Services